Rural SME Business Management

Human Aspects of Managerial Input

Rural SME Business Management

Human Aspects of Managerial Input

Peter L. Nuthall

CABI

CABI is a trading name of CAB International

CABI
Nosworthy Way
Wallingford
Oxfordshire OX10 8DE
UK

Tel: +44 (0)1491 832111
E-mail: info@cabi.org
Website: www.cabi.org

CABI
200 Portland Street
Boston
MA 02114
USA

T: +1 (617)682-9015
E-mail: cabi-nao@cabi.org

A catalogue record for this book is available from the British Library, London, UK.

ISBN-13: 9781800627222 (hardback)
9781800627239 (ePDF)
9781800627246 (ePub)

DOI: 10.1079/9781800627246.0000

Commissioning Editor: Ward Cooper
Editorial Assistant: Helen Elliott
Production Editor: Shankari Wilford

Typeset by Exeter Premedia Services Pvt Ltd, Chennai, India
Printed and bound in the USA by Integrated Books International, Dulles, Virginia

Contents

List of Tables

List of Figures

About the Author

Peter Nuthall has spent many years teaching and researching aspects of management. In addition he has developed and managed a team involved in producing and supporting computer-aided management systems used by large numbers of businesses. While most of his time has been at Lincoln University in Canterbury, New Zealand, he has also researched and/or taught at the University of Queensland, Purdue University (Indiana), the University of Kent and the University of Edinburgh. He has also worked for the UK Milk Marketing Board while based at the University of Nottingham (Sutton Bonnington). Many institutions involved in researching and teaching management have been visited to gather ideas. These include the Royal Agricultural College (Cirencester), Gatton College (Queensland), Swedish University of Agriculture Sciences (Uppsala), Wageningen University (the Netherlands), University of Guelph (Ontario), TEGASAC (Dublin), MAFF (UK), Cambridge University, Systems Support Centre (Aarhus, Denmark), Texas A&M University, and similar. He also has experience of production systems in a diverse range of situations including Russia, India, Fiji, Australia, New Zealand and the UK. Nuthall has published widely in scientific journals throughout the world and in monographs, as well as having items taken up by the popular press.

Acknowledgements

Many people have contributed to the material in this package. They include researchers around the world, managers similarly, consultants and textbook authors. Grateful thanks are offered to all these people.

But of special value was the contribution made by the case study managers and consultants who agreed to give full accounts of their lives and views on many issues. The true names have not been provided as some of the material is quite personal. Special gratitude is offered to these people.

And the professionals at the publishers have been a source of careful assistance and reminders. These have ensured the manuscript was checked with considerable care and provided in a reasonably timely manner. But above all, Ward Cooper, the Commissioning Editor, has shown faith in the concept and given valuable ideas for improvements. To all these people I offer a heartfelt thank you.

The publishers and authors who have provided permission to reprint some of their works are all gratefully acknowledged. The body of researchers and writers worldwide rely on each other for advancement and have a special bond. In particular ...

- Elsevier through Rightslink have provided permission to use some of the tables in Chapter 5 (Tables 5.1, 5.4, 5.5 and 5.6) and some of the commentary to these tables. This material was authored by the author of this book (2006).
- The Editor (Janice Osborn) of *World Agricultural Economics and Rural Sociology Abstracts*, published by CABI, has provided permission to reprint Fig. 1.2 (Human factors in business management) authored by G. Muggen (1969).
- Blackwell Publishing for permission to use Figs 3.2 and 3.4, which appeared in the *Australian Journal of Agricultural and Resource Economics* in an article titled 'Modelling the origins of managerial ability in agricultural production' authored by the author of this book (2009).
- Elsevier through Rightslink have provided permission to use the decision process table that appears in Chapter 4, 'Dynamic approaches' section, which was created by Ohlmer *et al.* (1998) in the article on decision-making processes published in the *Journal of Agricultural Economics*.
- The HayGroup (Ginny Flynn, Publisher) for permission to reprint details of the Kolb learning style types which appear in Chapter 2, 'Learning Style' section. Source: Kolb Learning Style Inventory. © Experience Based Learning Systems, Inc. All rights reserved. The content appears here with the specific permission of its publisher, Hay Group, Inc., 116 Huntington Ave., Boston, MA 02116, USA.

Finally, I am most appreciative of David Hollander's (retired Lincoln University photographer) efforts to find suitable photographs (Figs 4.2 and 10.1) to reinforce some of the messages. Figures 1.4, 1.5 and 1.7 are from iStock, courtesy of Abadonian, nikitos77 and Albert Karimov, respectively. Figure 1.6 is from Dreamstime.com courtesy of Patricia Hofmeester, and Fig. 6.3 is from Wikimedia Commons courtesy of Fahad Faisal. Besides the figure acknowledgements above, and those given with some of the figures in the text, the remainder – the majority – were provided by the author.

Research Background

Most of the evidence on which this book is based comes from a large number of studies published worldwide. Where appropriate, the detailed references to the study write-ups are given in the Bibliography at the back of the book, as are references that summarize all the research work discussed. This work encompasses many areas of business types (the 'domains'). In that small to medium businesses are frequently managed by an owner/manager, the result of investigating their operations reflects the decision systems and results from the skills of the manager. In corporate and multi-influencer businesses, the outcomes are the result of many people's skills and relationships. In these cases it is difficult to assess the influence of the individual contributors.

Accordingly, much of the research, but not all, that is quoted and used in this book is based on owner/operator situations given the clear link between the manager and outcomes. Thus the emphasis on 'small to medium' businesses. Furthermore, as an ideal set of information and data comes from primary production situations in the Western world where virtually all businesses are owner/operator, full use is made of this material as it provides clear links between management and outcomes. Full reference details are clearly provided to the research articles from this world of primary production, thus allowing the originals to be located.

1 Introduction

Abstract

It is stressed 'good management' as an entity is one of the four basic factors of production, which is a crucial input into successful SME business operations. This chapter covers the human details and structure giving rise to 'good management' as they are invariably not covered in any depth in management texts. To make better decisions it is important to understand the make-up and nature of managers and their characteristics which lead to good decision making. The top managers can be called 'experts' for they outperform colleagues successfully using techniques such as 'pattern matching'. Humans as business managers are defined largely by their personality, intelligence and possibly motivation. The facets of these characteristics give rise to a manager's 'phenotype', which can be changed to give better management with training and experience. The chapter ends by introducing some case study managers who appear throughout the book.

General Overview

Most production and related management economists refer to the production factors as land, labour and capital. While 'labour' might embody the managerial decision-making input as well as physical labour, it is clearer to separate management as a fourth factor of production. The decisions on how to use the production inputs and resources, and the implementation of the plans, are the responsibility of this fourth factor – management. In that the quality of the decisions gives rise to the success of the operation, this managerial skill is clearly absolutely critical to efficiency and profit. However, no texts and courses include the management factor in any depth. This book sets this situation to rights.

Texts on production economics cover the optimal allocation of resources. However, they largely assume that man is a rational being with near-perfect information. The reality is quite different. Managers are human. This means they react in an emotion-determined way. People observe the world around them and come to a conclusion about the current situation. Their mind, perhaps with the aid of calculations, comes to a decision over what actions should be taken. Thus, cues are observed that trigger action, or possibly inaction in some situations. This observation–decision–action process is something that varies with different individuals and needs to be understood if a manager is to improve the decisions aimed at achieving the firm's objectives.

The purpose of this book is to outline the human components of what makes a person, and why a manager acts in particular ways. This understanding is essential in assisting business operators to improve their management, and thus attain their objectives. This assumes that improvement is indeed possible using various techniques. Evidence pointing in this direction will be reviewed. As understanding provides wisdom, the emphasis is on looking at theories and their application in contrast to rote-learning rules and solutions. These seldom cover the myriad of situations possible, thus leading to misinterpretations and mistakes. Armed with knowledge on the parameters that determine how an individual reacts, and their relationships, it is then possible to treat each unique case on its merits.

There is ample evidence that managers vary markedly in their skills. Profit and efficiency data from businesses in similar environments makes this clear. For example, studies that relate individual businesses' position relative to their peers show the average technical efficiency can be as low as 36% (a Swedish study) and as high as 85% (a Pakistani case). Of course, these studies assume that all the managers have the same objective. In reality, some will be happy to be less technically efficient if it means, for example, their average production is quite

DOI: 10.1079/9781800627246.0001

stable. Similar ranges in efficiency exist when using economic outcomes as the yardstick. One study for Brazilian businesses put their average efficiency at 13%, in contrast to a US dairy business study giving an average efficiency of 70% (Dhungana, 2000).

These efficiency studies determine which managers are producing the greatest output for given inputs, and then relate other managers against these producers as benchmarks. If the average efficiency was 100%, this would mean all managers are producing the same ratio of output to inputs. As a comparative measure, this does not necessarily mean that the 'efficient' managers are in fact efficient in an absolute sense. No doubt they could increase their efficiency through even better decision making, using perhaps the latest technologies. This can only be judged if, for example, comparisons with perfectly managed demonstration businesses are possible.

The managers of businesses that are the most efficient in a sample can be called 'experts'. Studies of experts show they have particular attributes. These include:

- their expertise is restricted to a limited domain of operation;
- being good at clearly defining a problem;
- accurately observing relevant cues and the importance of each;
- automatically perceiving meaningful patterns;
- coming up with solutions almost instantly;
- having superior short- and long-term memories;
- observing and characterizing a problem in terms of its basic structure;
- spending considerable time quantifying and analysing problems, especially if the particular problem has not been faced before, thus adding to their understanding and store of knowledge; and
- clear self-monitoring abilities leading to improvement (i.e. good and objective self-criticism).

Relative to novices, experts know what to observe, do so quickly and accurately, and use their experience to provide a solution that is appropriate. If the problem has never been experienced before, their superior processing systems work out the solution, and then store this for future reference. With time, they become real masters. If any one of the characteristics of an expert is not present, the person will not attain the true expert classification, though there will be degrees of expertise. The important questions relate to the characteristics required to become an expert, and how this status can be acquired.

It was pointed out experts are quick with their judgement and decision. This is most likely due to pattern matching. This refers to having the pattern of a problem, and its solution, stored in memory so that when the expert observes the values of the critical parameters, this set of data is sent to memory to find a match with the stored information. When found, the solution is readily available. These patterns might be visual (a picture of some kind), or abstract (lists of benchmark data perhaps). For example, if you see a bush with flowers on it of a certain nature, if you have seen the bush before and its image is stored in your memory, it is immediately recognized as, say, a rhododendron. Analysis of some kind is not necessary. However, if the bush is not recognizable, research is required. This might be by referring to a book, or perhaps a recognized expert. Next time, you have this newly acquired pattern stored in memory, so an immediate recognition occurs.

An important question concerns the personal qualities that are necessary to become, in this case, an expert business manager. While training and experience can make up for a lack of inherent ability, a good measure of both is probably highly beneficial. Being an expert enables appropriate decisions in good time. Good managers seldom have to spend large chunks of time to sort out a problem for their mental systems have the right answer stored. This ability to make a decision quickly is sometimes referred to as 'intuition', or sometimes 'tacit knowledge'. This sounds like a mysterious quality that only some managers have. In fact, it probably relates to having the right attributes, experience and training. Thus, decision ability from intuition is probably a learnt attribute that we all have to a greater, or lesser, extent. As it is not appropriate in many cases to spend a long time researching a problem or opportunity, developing this intuition is an important aspect of being a good manager. Similarly, knowing when your intuition is likely to be incorrect is also obviously an important attribute. In such cases, formal study, research and analysis are required. This usually leads to an enhancement of the manager's intuition.

Skill and intuition must cover a wide range of areas for successful management. Any business production system, of food and fibre production, as an example, involves an extremely wide range of necessary skills. Frequently production involves:

- a wide range of resource inputs including rainfall and climate;
- biological systems, harvesting and machinery in general, engineering (buildings, structures, designs ...);
- labour and personalities;
- markets, finance and economics; and
- politics and the resulting impacts on the rules and regulations that must be complied with.

This very wide range of subjects covers everything from physics and chemistry through to biology and psychology in that the people involved operate within the bounds of their human characteristics. The excellent manager will have a reasonable understanding of all these areas.

Nevertheless, equally, if not more, important are the management skills that a manager brings to the job, which, in turn, lead to the decisions made and implemented. Thus, a manager must know how to use facts and figures through being skilled in:

- *understanding the technology* and what lies behind it (meteorology, inputs and their basic characteristics, raw materials and their relationships, outcomes, the sciences – biology, physics, etc.);
- *observation and recording* (production machine conditions through to international markets);
- *planning* (risk management, cash flows, job priorities, time management, economic principles, etc.);
- *anticipation* (possible outcomes and their chances);
- *people skills* (labour management, network maintenance, negotiations, etc.); and
- *personality management* (stress management, motivation, objectives, and so on).

Acquiring abilities in all these areas is dependent on the basic attributes of a potential manager, and the opportunities for training and experience that are made use of. Some will be good at a sub-sample of the attributes, and some will be rounded with a complete package of excellent skills. The critical question relates to how a manager might acquire such a fully rounded set of attributes.

It appears that a human being is defined by two, possibly three, basic sets of factors:

- the first is a person's personality;
- the second their intelligence; and
- the third is their motivation, although some researchers believe motivation arises from their personality and intelligence.

Personality is made up of sub-factors such as extroversion and anxiety levels, and intelligence is similarly made up of components such as memory and reasoning. In each segment, an individual will have a defined make-up, leading to the unique whole. Some will be good managers, others not.

When observing the attributes of a manager, you observe what is called his or her 'phenotype'. In contrast, the package that a person is born with is referred to as the 'genotype'. This is defined by the inherited genes passed on 50–50 by the parents. But the genes only define the person's building blocks, which then interact with the environment and experience that a person is exposed to. The sum of the genotype and environmental experiences gives rise to the observable phenotype.

The question is what phenotype is appropriate for good management, and how is this attained? Clearly the genotype cannot be altered, nor can a person's early experiences – this is now history. However, some phenotypes will more than likely be capable of being modified through the correct training and experiential exposure. For a manager to improve his or her skills, it is clearly important to discover what training will in fact work. This assumes phenotypes are alterable, and fortunately there is evidence this is indeed the case. Understanding phenotypes and managerial skill is also important in setting up the correct conditions right from an early age to ensure that good managers are created. Clearly, the approach of providing an appropriate set of experiences and training in early life is preferable to trying to improve skills in later life, though both possibilities must be followed.

People, and managers, are seldom totally consistent in their actions through time. This might be because they have re-evaluated their objectives, but more likely simply because they are human. While an individual has fixed traits defining her or his personality and intelligence, how these give rise to decisions made on any 1 day can be variable. You will be aware that your emotions and state of mind vary from time to time in response to events and activities. This is an expression of personality, with some people varying more than others as defined by their phenotype. The consequence is that the decisions made will not always be consistent, even given the same circumstances, particularly for people with an anxious personality. Effectively, a person's state varies, so the astute observer must be aware of this and not conclude on a person's managerial ability based on a restricted number of observations.

While learning the basics of what gives rise to a person's actions is valuable in itself, the important aspect of this understanding is considering how managerial skill might be improved. Most managers improve with time as they take in the lessons of experience, but the extent of improvement probably depends on their personality and intelligence. This situation has been expressed in Fig. 1.1.

At any given point in time, a manager embodies a particular skill level that is dependent on their past environment, intelligence, age and personality. Also relevant are the manager's objectives in that, for example, she, or he, might not consider efficiency as being relevant. Also possibly relevant is what is known as a manager's 'locus of control'. This reflects the attitude to how much control over the business's destiny the manager believes s/he has. Some believe, for example, that outcomes are largely determined by random events such as the weather as well as the world markets and, therefore, managerial skill is not that important.

Then, through time, change in skill is dependent on the training and its effectiveness. Success may well depend on the type of approach the manager prefers relative to the structure of any training offered (preferred learning style). Precursors to formal training, if any, are the manager's desire to involve her/himself in training, and then her/his potential to actually absorb and learn from the training. Often, the training is totally informal and relies on the manager's reading, observation and ability to learn from mistakes and experiences. Understanding the personality and intelligence of a manager helps understand whether s/he will improve, and the best way of undertaking this improvement.

Of course, any formal training comes at a cost. There might be formal tuition fees, and certainly there is a time commitment with the attendant opportunity cost (income that could be obtained from the time used in the next best alternative). There may well be an emotional

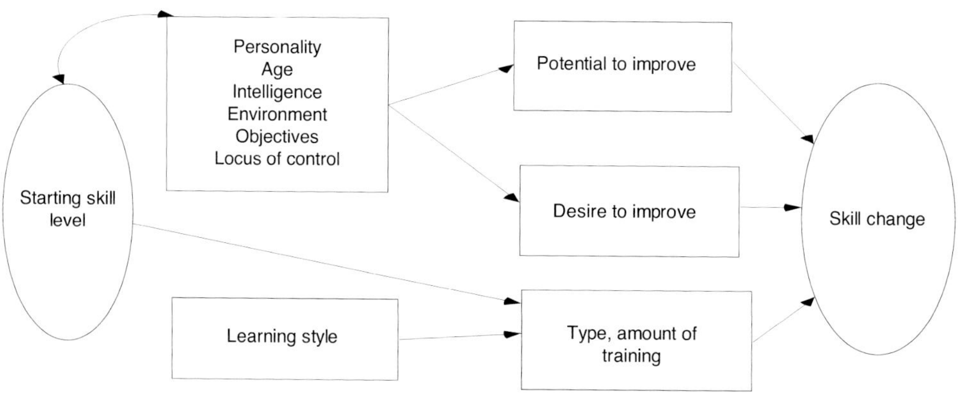

Fig. 1.1. A transactional model of improving managerial skill.

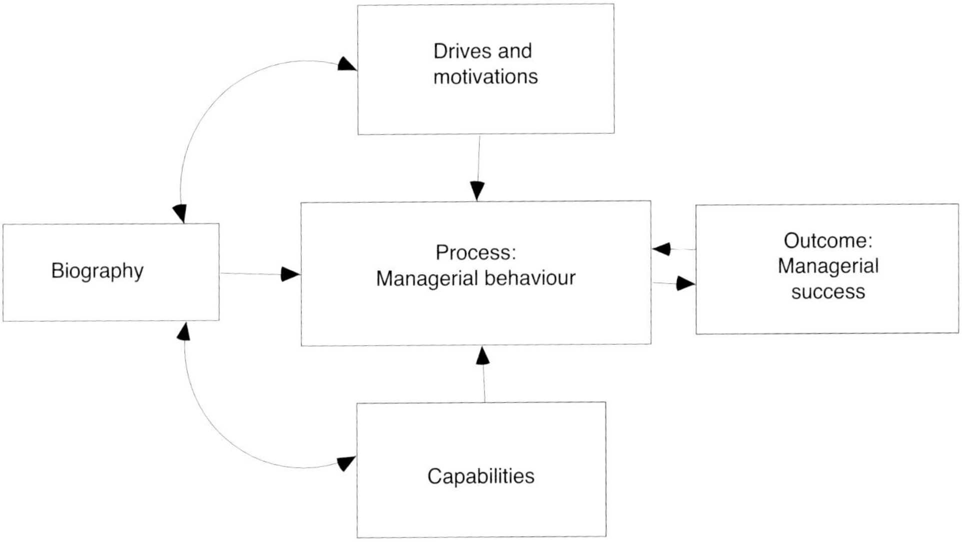

Fig. 1.2. Human factors in business management. (From Muggen, 1969. See the Acknowledgements, p. xv, for reprint permission details.)

cost as well in that 'going back to school' can be traumatic. Some managers will not be prepared to accept the discipline involved in attending evening classes after a day of mental, and perhaps, physical activity, and commit time to working through exercises. Besides the emotional and discipline issues, the decision to take up formal improvement also involves economic issues. The decision is very much a matter of assessing the costs relative to the gains from the improved management. However, it does not take much improvement to cover the time cost, particularly where the training improves the manager's intuition that will have ongoing value up to retirement, and probably beyond.

Historically, business researchers and writers have commented on the importance of managerial skill, but this aspect of production efficiency is seldom highlighted, and the level of research funds devoted to the area is quite minimal. This situation needs to change as any manager is clearly the key to combining resources appropriately to achieve the business goals.

In considering food and fibre production as an example domain, early texts, for example Case and Johnson (1953), note the importance of the manager, and, from time to time, research on the management process has been published. Research examples include Johnson *et al.* (1961), who studied Midwestern managers in the USA with respect to how they operated; and Ohlmer *et al.* (1998), who similarly studied Swedish managers and concluded that their decision process was dynamic with constant adjustments and reviews.

Furthermore, there has been a range of studies over the years that have related human factors with success and efficiency. The factors used in the analyses have ranged from the manager's age and education through to psychological measures. Early work considering the factors was reviewed by Muggen (1969), who concluded that there were 61 variables correlated with success. Figure 1.2 gives his summary of the situation. The variables included socio-economic status, education, motivation, vocabulary, technical knowledge and many more.

Other studies looked at:

- human capital relative to business size and growth (Summer and Lieby, 1987), human capital and SME growth (Shoubaki *et al.*, 2020);
- psychological type and financial success (Jose and Crumly, 1993; Farrington, 2012);

- thinking styles and financial characteristics (Howard *et al.*, 1997; Day *et al.*, 2006);
- experience and production efficiency (Wilson *et al.*, 1998; Xin Ding, 2014);
- personality and intelligence relative, in particular, to environmental behaviour (Austin *et al.*, 1998; Brick and Lewis, 2014);
- defining an operator's management capacity (Rougour *et al.*, 1998; Izadi *et al.*, 2020);
- measuring managerial efficiency using personological variables (Warren *et al.*, 1974; Tripp *et al.*, 2000; Robie *et al.*, 2008);
- personality measure and managerial ability (Young and Walters, 2002; Myszkowski *et al.*, 2015);
- predicting management performance using manager decision-making profiles (Solano *et al.*, 2006; Montequin *et al.*, 2015); and
- the importance of training on ability improvement (Ellinger and Ellinger, 2014; Xayavong *et al.*, 2015).

These studies show that management skill is indeed correlated with many factors that together define a decision maker's managerial ability, or what can be called 'human capital'. The limited amount of research shows there is still a general lack of recognition of the importance of studying and understanding managerial ability. Furthermore, the studies have not provided a basic theory and understanding of the creation of a manager's skill, or lack of it, nor of how an individual's skill level can be improved.

The following chapters bring together this, and new, work as well as developments and theories from other disciplines, including psychology, to provide a well-structured set of ideas for understanding managerial ability. Chapter 2 contains outlines of the characteristics of a manager that defines her/his modus operandi, and how these characteristics might be measured using psychometric tests. This is a person-centred approach in contrast to the use of production parameters that are the outcomes of the decisions made.

Chapter 3 looks at the relationships between the factors that define a manager. What is the influence of each in creating the success, or otherwise, of the managerial ability? Clearly knowing the importance of the precursors to ability enables recognizing the areas that might be improved with reward. Effectively, the discussion covers the 'origins of ability'.

Also important is an understanding of the processes involved in comparing alternative decisions, and, therefore, actually making decisions. Consequently, Chapter 4 is devoted to reviewing the possible decision criteria and processes used. These criteria range from the concept of maximizing a general measure of output – utility – through to the simple approach of producing a minimum requirement of an output such as leisure time based on intuitive feelings.

Having covered the factors giving rise to ability, Chapter 5 contains an outline of the competencies necessary for successful management. Obvious areas are skills such as accurate observation, and an ability to forecast or anticipate outcomes following specified decisions. These and other competencies are reviewed using managers' views.

In that managers are seldom purely rational 'machines' in their decision making, there is an implication that their decisions can often be biased – that is, incorrect or wrong relative to their objectives. Thus, Chapter 6 outlines the common errors, or biases, and, therefore, highlights areas that require special attention when improving ability.

It has already been noted managers make many decisions using their intuition. Chapter 7 focuses on this topic by outlining a quantifiable model for use in assessing the important factors influencing a manager's intuitive skills. The results of using survey information to quantify the model are discussed with conclusions on the important aspects in improving intuition highlighted.

While the emphasis in the book is on managerial ability, a manager's personal characteristics also have an impact on many other important issues of a business and its management. Chapter 8 contains four case studies exploring four example decision areas involving growth, ownership structures, succession and manager anxiety. In each case a model is developed and used as a basis for analysing manager data to show the impact of, particularly, a manager's personality and objectives.

Most SME managers worldwide are owner/operators in various forms. The essence of this is that usually a family is intimately involved in the business 'life' and decisions, with the manager and his or her family living and working close to the business. consequently any study of a manager/owner for an SME must also involve a study of the family and their influence on the objectives and decisions taken. Chapter 9 considers these influences and also the factors that give rise to the objectives held by the manager–household complex, for business success must revolve around both having the objective set that is right for the family and having systems that enhance the attainment of these objectives.

Chapter 10 then reviews whether success is possible from using manager characteristic improvement programmes. If, in fact, the precursors of ability are fixed for all time following birth and early experiences, there would be no point in training and development schedules. This is not, however, the case.

Finally, the various threads of the arguments are brought together to provide conclusions.

In summary, this book provides an understanding of a manager's characteristics and why a particular person reacts in a defined way when faced with a range of situations. The important features relate to the nuances and attributes of a person's expressed personality and intelligence together with how the person's persona impacts on such things as their learning style, motivation, objectives and belief in their control of their management destinies. As the majority of small to medium business situations involve an owner/manager arrangement, the family is also an influence on the outcomes and their efficiency.

Besides using casual observation to assess a manager's characteristics, this book also provides written tests to gauge business-related personality, intelligence and the locus of control. Benchmark data for a sample of managers is also provided. It also provides the results of surveys to assess what managers consider as the important skills associated with successful management. Question sets are also provided to assess a person's objectives and the important physical and managerial aspects of a business, for there is an intimate association between a manager, and his family, and the business itself (as in Fig. 1.3) which can influence the skills required.

The contents of the book do not cover all skills that might be important as there are many texts available on things like negotiation skills, labour management, listening skills, production economics and related computer packages

Fig. 1.3. Being a rural manager confers many benefits. The beauty and remarkableness of the country confers constant wonderment.

covering, for example, mathematical programming and optimal seeking systems, regression analysis, cost–benefit analysis, and the like.

However, what have been included are the comments and thoughts of two managers whose situations have been used to demonstrate the various factors raised in the book. In addition, for the chapter on the methods of improving managerial skills (Chapter 10), the experience of two business consultants has been explored to obtain their views. Details of these two consultants are provided in the chapter on the skills required (Chapter 5) as consultants also have their views on the important attributes. But it is appropriate to introduce the two case study managers at this point.

The Case Study Managers

Introducing Margrave

The first case study manager has suggested he be known as Margrave. Perhaps his ancestors had connections to the Margraves of olden times for there are several castles and other sites of this name in Europe and Britain. This only emphasizes that Margrave has had connections to businesses for centuries through his ancestors that he has been able to trace back to 1066, with every generation being a food and fibre producer in some form. Therefore, you have to say production business is in his blood. Margrave manages what in Western terms would be called a significant operation with eight full-time employees, though it also involves forestry (see Fig. 1.4 as an example of a forestry operation) that requires less-than-constant attention with activities such as pruning being able to be conducted at times that fit in with other activities. However, he was not always involved in such a large investment for he first started business in a small way before progressing to another business, and finally ending up on his present operation. Margrave's operation involves food and fibre production with neighbouring vineyards. The business has faced difficult conditions at times.

Total stock involves some 39,000 units, where a unit is defined as the food needed to support a 50 kg animal (made up of a mixture of animal types). The large numbers of animals purchased in the summer enable greater flexibility in the system, so if the summer has

Fig. 1.4. Each season brings new hope. A scene typical of what inspires Margrave in his fascination with production systems. (Courtesy of Abadonian, iStock.)

particularly bad rainfall, the numbers purchased can be reduced. However, in normal times, if there is such a thing as a normal season, life can proceed smoothly but such situations seldom exist given all the risk and uncertainty in production conditions and markets.

Margrave has been a manager for 14 years (Fig. 1.5), following 5 years working in other businesses. Margrave was educated in a small primary school but was sent to an urban boarding school for his secondary education, which he notes 'was an experience, but I didn't work much, relying on natural ability to get me through'. Margrave was not really exposed to the decision making with his family keeping discussions about business and family matters to themselves, something that was not uncommon in earlier times: 'Private matters were not

Fig. 1.5. Large stores of products can be somewhat daunting, but not for Margrave. Piles of resource needing grading and sorting. (Courtesy of nikitos77, iStock.)

Fig. 1.6. Modern milk production and processing is capital intensive. A scene typical of Hanna and Hank's investment. (Courtesy of Patricia Hofmeester, Dreamstime.com.)

discussed with me.' While Margrave did help with physical jobs, he notes 'It wasn't until I was 16 that I learnt much about the production processes.' He left school about this time.

Introducing Hank

The second case study is also about a producer who operates a totally different system, and also has a totally different background. He is involved in operating two milk production systems. His early experiences were on a family farm, but his initial working life was spent in the banking world. This very successful SME manager wants to be called Hank despite not being originally from the Netherlands. In fact, he was born and bred on the west coast of the USA. Margrave, on the other hand, was born and bred in New Zealand. Hank talks about his ancestors and notes that they were involved in farming in the Midwest of the USA. One grandmother was so attached to the land that when she died on the west coast she wanted her body to be shipped back to the Midwest for burial. This strong attachment is not at all uncommon. A manager's heart and soul are often tied up in the life-giving material upon which we all gaze from time to time.

Hank and his wife, whom we will call Hanna, started managing on their own account in 1987, and ever since then they have grown their investment by acquiring more resources and indeed only recently purchased an additional operation for milk production. As noted, Hank was born in the USA, whereas Hanna is a New Zealander. Hank's father was both a dairy farmer and golf course owner, so he was probably an entrepreneur. Hank worked on the farm as a child and has very fond memories of his father, with many long chats about decision making. They also had special times together travelling to other operations to inspect their systems. They were, clearly, the greatest of buddies. However, Hank's father died when he was just 16, and his much older sisters became important in his life. Yet, the seeds of a passion for food and fibre production had been well and truly sown. Hank eventually ended up at a major land grant university, studying management and finances that led him into the banking world where he spent his time assessing loan requests from managers and over-seeing their use of the money. He enjoyed this job, and quickly rose to become a vice-president at the age of 33 years. Part of his education involved a visit to New Zealand on a university-exchange programme. He is not sure why he applied to be one of the first on the programme, but he ended up being selected and travelled to Lincoln University, which is where he met his wife-to-be. Later, Hanna travelled to Hank's American university on the reverse exchange programme.

Hank maintained contacts in New Zealand, and even bought assets from a distance. One day a friend in New Zealand suggested that he apply for a share-manager's position, and also buy additional resources as prices had reached rock bottom with major changes in government policy. As Hank could not see himself working the rest of his days in a bank, and having observed how his older colleagues were locked into the system with pension plans that meant they could not leave without major financial sacrifice, he applied for the job. Fortunately Hank was successful though he does suspect, given the poor state of the business world at this time, he was the only applicant! This start was in part possible due to the assets Hank had in New Zealand that had been managed by 'remote control'. This background led to Hank and Hanna eventually setting up their own business and the significant holdings they now manage (Fig. 1.6).

Fig. 1.7. Skill levels must be high to obtain a return from intensive and highly leveraged systems. (Courtesy of Albert Karimov, iStock.)

Hanna and Hank now operate two businesses using various ownership structures. The total production system is run with eight labour units using various share agreements covering both the costs and profits. If there is a downturn, everyone helps take the brunt, and vice versa, though there is a baseline system to ensure the share owners have at least a minimum income.

With the development of improved systems and management, production and profit have steadily improved (maintaining quality is critical, Fig. 1.7).

Hanna and Hank started with very little, and are proof that careful, but also excellent, management can give rise to a major investment with high economic production. Their human story is worth listening to.

As appropriate, the stories of the two case study managers will be introduced and discussed in selected chapters with respect to the particular issues raised. They provide both concrete and human evidence of the issues presented.

Besides these case studies this book also contains the results of a considerable number of research studies into all things human as they relate to management and the human factor. These studies are, primarily, of food and fibre businesses in New Zealand as this type of SME provides an ideal way of clearly isolating the impacts of a manager and her or his characteristics. This is due to the businesses being largely owner/operator-managed entities, making it much easier to isolate the impact of the manager due to the relatively simple control structures. The businesses cover a range of sizes with a net asset base of around US$4,000,000 overall, with 34% of the managers having tertiary education and another 30% having 4–5 years of secondary education (Fig. 1.7). Overall, the average grade obtained by the managers over their last year of formal education was 63%.

2 What Defines a Manager's Ability?

Abstract
Defining managerial ability involves understanding the building blocks, which include personality, intelligence, planning and replanning ability, memory and analytical skill, among other human characteristics. The details and properties of these issues are discussed. Personality (or 'management style' in reality) has five core components which are fully described. Similarly, the factors involved in intelligence are covered though the direct correlation between intelligence scores and managerial ability is not high. However, an example of a management 'aptitude' test is provided which more likely covers the important components of management appropriately. Tests for both management style and aptitude are provided and discussed. In addition, a test for a manager's attitude to risk is provided and similarly discussed. As becoming a good manager requires heaps of learning and observation tests, a manager's learning approach is also covered. Finally, it is stressed understanding a business's objectives is critical for good management for they provide the yardstick for choosing between alternatives.

Introduction

Anyone can be a manager – but they may not be particularly good at achieving their objectives. The desire to be a manager is relevant in success. Some managers perhaps acquire their status due to tradition and the handing over of assets rather than a keen desire to make a career from managing production and marketing. On the other hand, some who want to be business owners and managers find it is impossible due to the resources required. Whatever the case, an ability to take and accept the responsibility of making and carrying out decisions is an important precursor, but whether a person can in fact make good decisions relevant to a particular business (for each is unique) depends on whether they have the required attributes and experience. Abilities such as making sure that the jobs are carried out in a timely manner (e.g. attending to a market before a competitor has moved into it; getting supplies delivered before it is too late to complete the job; marketing the product before the prices drop) are crucial and probably relate to a person's degree of conscientiousness as well as an understanding of the systems and marketing principles involved. Conscientiousness is a personality trait. Similar examples exist for the other attributes, so a manager's personality impacts on their likely success as a manager.

Management also involves holding in memory a large store of recallable information – thus a good memory is a valuable attribute. It is also important to be able to sort the grain from the chaff when considering, for example, inputs being offered by a myriad of salespeople. In primary production fertilizer choice is a prime example. Getting it right requires clear and logical thinking, using knowledge of the relevant technology such as, in this example, soils, fertilizers and nutrient requirements. These two examples relate to a manager's level of intelligence, so, as you would expect, reasonable intelligence is a requirement for good management. This does not necessarily mean high achievement in formal education as there are many examples of excellent managers who have had little schooling. On the other hand, there are an increasing number of managers in the developed world holding tertiary qualifications.

Even given the right personality and good intelligence, a new manager will struggle to get it right if lessons have not been learnt from experience. Managing a business is a very practical occupation where a good knowledge of practice is important. A good understanding of systems and outcomes is most unlikely without considerable experience and lessons from past mistakes. Being involved in situations that cover the problems and conditions likely impacting on a business is important, as is an ability to self-analyse

DOI: 10.1079/9781800627246.0002

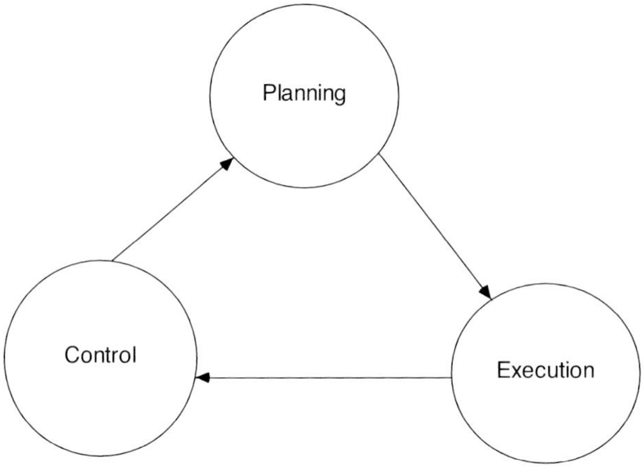

Fig. 2.1. Responsibilities of management.

and benefit from the experiences. Learning from experience is probably dependent on good intelligence and observation systems. It should also be noted that while some people cannot formally analyse what they have learnt, their intuition may be finely tuned, having inherently picked up the lessons on offer.

A manager must be very clear over what his/her responsibilities are. Figure 2.1 outlines these in a general sense. Action and decisions always start with planning, the conclusion from which might be to actually take no action. Thus, a review, for example, of whether to market a maturing wine might conclude that it is not yet ready. The plan is 'wait and review'. Other plans where action is required then move on to the execution phase. The plan must be put into effect. This involves, potentially, the myriad of actions necessary in getting a plan completed. Manufacturing kitchen appliances, say, requires:

- ordering and receiving all the steel and timber, fuel and other supplies;
- checking the operational status of the equipment required;
- getting the labour organized; and, finally,
- actually getting the components made and fabricated in the correct order, then painted perhaps.

Prior to this, of course, the production area had to be appropriately prepared (through altering various tool settings and dies ...).

Once the plans have been activated, the next phase involves 'control':

- the outcomes must be observed to ensure that the expectations occur; and
- where a variation is observed, replanning may be appropriate; and
- so the cycle repeats itself.

For example, as the coverings are applied they might turn out to be incompatible with some of the materials, requiring respraying or, perhaps, altering the publicity brochures.

This whole planning–execution–control cycle is a very dynamic operation, with constant observation and replanning leading to further execution. Particularly in export industries including agriculture, plans are seldom completed as originally expected due to the constant vagaries of the weather, markets and other risk factors such as the availability of suitable labour.

A successful manager must recognize this dynamic process, and always be prepared to reanalyse and change plans and actions. This is what makes it a wonderful challenge in which only the skilled really succeed. It is easy to see why an appropriate personality and good intelligence are highly desirable. However, any manager can have a particular phenotype (make-up) that may not be what is best for good management, so in these cases it is important to recognize this and make efforts to make as many modifications as possible.

This improvement process is helped if managers are able to classify themselves, or to seek assistance from a professional, to determine where they might change to improve their management. Absorbing such changes to provide permanent improvement is desirable, given that most of the planning–execution–control process is carried out intuitively, in contrast to a formal paper-based operation. Whether, and when, to start up a special advertising programme, for example, will be an intuitive decision based on careful observation of the markets, competitors' actions, inventory, inflation rate tendencies and cash supplies.

Classification of managers into skill groups can also be intuitive, or more formally based on observing outcomes. The ultimate criterion is the 'end-of-year profit' relative to peers in similar businesses, and also achievement in other objectives besides cash (reducing global warming outputs, perhaps). Alternatively, it is possible to administer written tests, the results of which can be assessed against benchmarks. The reasons for tests include providing a knowledge of the manager's current ability, and thus provide a baseline from which changes can be observed. This baseline test also helps indicate whether improvement is a real possibility.

If the manager is already exhibiting all the right characteristics, then only fine-tuning will be possible, perhaps in conjunction with a professional consultant. The tests can also suggest the particular areas that might be worked on for improvement. For example, a specific manager might exhibit a high level of anxiety, resulting in apparent stress and inappropriate decisions that, perhaps, are designed to remove the chance of risky outcomes. If this proves to be the case, a programme designed to dampen the anxiety could be appropriate. Similarly, it is useful to know a manager's strengths so that they can be exploited. For example, a strongly extroverted person may well be able to excellently manage a committed work team, enabling, say, intensive production systems using a large workforce for good reward.

Tests such as those discussed are generally called psychometric tests in that they are designed to explore a person's psychology – that is, to find out the characteristics that determine how a person will react under a range of circumstances. The study of psychology is the study of people and how they react to their environment. This is exactly what we are talking about – how managers react following observed cues from around their business, markets and the wider community. Psychologists have developed a large number of tests to explore a wide range of personal characteristics.

Most people have experienced an intelligence test at some stage in their life. These tests are mainly designed for the general public, whereas in specific occupations and industries there may be specific attributes that are important. The development of suitable tests for assessing, say, primary producers is only in its infancy, with limited experience available. Later in this chapter some of the tests that have been developed are provided together with benchmark data. In a general business sense, psychometric tests have been used for many years, but many are not suitable for specific production and tertiary operations as the situations are quite different. However, some of these general tests continue to be used in some situations, indicating they have been valuable. Many people, for example, have experienced at least one of these when thinking about the occupation they would like to spend their life in.

As noted, the two critical aspects of a person's management are their personality and intelligence. The next sections in this chapter cover the important aspects of these two broad features of a person and introduce tests that can be used to assess them. The 'personality' test is actually slightly different from a standard test in that it is called a 'management-style' test, as this is the aspect of personality that we are interested in. While modern personality theory suggests that people have five basic traits, the management-style test looks at six factors which include the basic traits. Similarly, the 'intelligence' test outlined is referred to as an 'aptitude' test, for it is a person's managerial aptitude that is important. Again, the components of this test relate to the intelligence aspects that are likely to be important in business management.

Other tests that can be important include the 'locus of control', which reflects a manager's belief in how much control s/he believes s/he has over outcomes. Assuming control is indeed possible, a belief that you can in fact influence outcomes is obviously important. A test for the 'locus of control' is provided in Chapter 3. Also relevant is a manager's preferred way to learn new skills, in that some people do have abstracting abilities and can learn from books in contrast to needing practical demonstrations and experience. A test for learning style is available and will be discussed.

As some business can be very risky, it is important for a manager/owner to be fully aware of their attitude to risky decisions. Thus, tests to determine risk attitude are introduced. With intensive effort it is possible to alter these attitudes with some benefit. A manager's objectives also relate to this question, therefore a brief discussion on objectives is also included, which is expanded in a later chapter (Chapter 3). Similarly, motivation can be an important factor related to objectives, so some comments are provided on this subject. Finally, a test called 'emotional intelligence' is introduced because it is important in urban business, but its application in solely rural production businesses is probably much less significant as it relates to constant dealings with other people including large numbers of customers.

Personality and Its Links to Management

While many ideas on personality have been suggested over the years, and this debate still continues, many psychological researchers consider the five-trait model as being correct. One text that describes the model is provided by Matthews and Deary (1998) with references to many of the basic research articles giving rise to the five-factor model. While each base factor can be considered made up of many facets, the core traits are:

- 'openness';
- 'conscientiousness';
- 'extroversion';
- 'agreeableness'; and
- 'neuroticism'.

It will be noted the first letter of each makes up the word OCEAN. While the word neuroticism is commonly used, in a management sense probably a better word is 'anxiety'. Neuroticism has connotations of the extreme ends of the spectrum. Although we all have a particular level of this trait, it is likely to be far from the extremes and, therefore, make us anxious rather than neurotic in the everyday use of the word.

The openness trait is expressed through a person's attitude to new ideas. A particularly open person will be daring, liberal and somewhat original in their thinking. In contrast, a person who scores poorly on the openness scale will be conservative, unadventurous and conventional. It is clear that it is desirable for managers to have the right degree of openness to suit the situation they operate under. In an environment where change and opportunity are available, a high degree of openness will enable capturing the potential benefits of new products, new methods and systems. On the other hand, where the markets, technology and political conditions are relatively constant, lack of openness is not much of a problem. On balance, provided an 'open' person does not get bored, a manager who is very 'open' is more likely to succeed, given that new products, ideas and opportunities are always likely at some stage in a manager's history of work. In real life, managers will exhibit a full spectrum of degrees of openness. A manager's suitableness for each environmental situation will similarly follow a full range.

Conscientiousness is clearly an important trait for successful management. Someone who has a high rating on this trait will be careful, reliable and takes responsibility seriously. Such people can generally be relied upon, and when a task is agreed you can be sure it will get done. In contrast, someone who exhibits the other end of this trait's scale will be somewhat careless, undependable and even negligent. While, no doubt, there are some successful managers who show little of this trait, possibly through good luck, generally a good level of conscientiousness will be associated with successful managers. On the other hand, too much conscientiousness could be counterproductive, particularly when associated with a low level of openness. Being extremely careful, with little daring, could well stifle good management. This idea starts to indicate just how complex we are. Once two traits are considered, each of which can have a wide range of levels, a specific combination will have a specific relevance to managerial skill. Once all five traits are considered, it is highly likely that any one manager will be unique relative to all other managers. Tests, as discussed below, enable sorting out where a manager lies in the full spectrum of possibilities.

Extroversion is a trait which most people recognize quickly. An extrovert is sociable, talkative and frequently spontaneous. You probably know people like these, and perhaps you are an extrovert as shown by how you seem to be energized when talking to a group of people. Extroverts will enjoy both management and social events in contrast to the opposite personality, referred to as an 'introvert'. Such people are probably shy and find mixing with many people difficult and not to be courted. Thus, an introvert will be retiring, quiet and somewhat inhibited. As a factor in good managerial skill, a degree of extroversion is necessary in that it is important to interact easily with people when looking for information and help. An introvert will be backward in seeking help, be it in a professional sense or in a personal help situation. Also important is how well a person manages staff and contractors. An extrovert will generally find it easier to express requirements and feelings, both being important aspects in running a successful team. In a nutshell, extroversion is probably a factor in leadership skills, though other attributes are no doubt also relevant. On the other hand, for businesses where a big team is not required, a degree of introversion is probably quite acceptable for good management.

Agreeableness is the fourth basic trait. A 'basic trait' is a distinct feature that is independent of all other traits. The research that suggests these five basic traits considered correlations between a very wide range of personality features and discovered that many of them were correlated. If the correlated groups are separated, the OCEAN traits are left as independent. A person classified as being 'agreeable' is good-natured, soft-hearted and somewhat selfless. Generally these people might be called benign and seldom get angry or overly excited about issues. At the other end of the spectrum, a 'non-agreeable' person will be rather irritable, and certainly ruthless as well as being somewhat selfish. It will be noted that these characteristics are somewhat disparate, but research has indeed found that they tend to be highly correlated. In a management sense, a certain degree of 'agreeableness' is probably helpful in that a calm and accepting nature helps handle the ups and downs of working in an environment over which there is little control. When the hailstorm wrecks the buildings and contents just before shipping out a batch of product, it takes a very agreeable person to maintain an objective view of the situation. Similarly, when the employee reports he has driven an important machine into a post, destroying the structure and denting the equipment, an agreeable outlook is most helpful.

On the other hand, there are situations where it does not pay to be agreeable. Sometimes getting a little irate with a dealer, or perhaps a contractor, when the job is not being done as required, can achieve some much-needed action at the right time. Thus, a personality that is on the agreeable side of the middle line is helpful, provided assertiveness can be produced when some positive action is clearly necessary.

Finally, there is the 'anxiety' trait (or neuroticism as defined by psychologists). An anxious person will worry, be somewhat vulnerable to less-than-favourable outcomes in that they will easily become emotionally upset and also exhibit insecurity. That is, they will tend to be unsure of whether they made the correct decision and often worry endlessly over the decision after the event, even though they know full well the die is cast and nothing further can be done about the situation. Most people will have at least one acquaintance that can be called anxious. In the extreme, anxiety can lead to physical illness and other undesirable characteristics like endlessly going back to check, for example, whether the stable door has been successfully locked.

A non-anxious person will, on the other hand, tend to be calm, resilient in contrast to vulnerable and feel relatively secure in the decisions made. In that anxiety can lead to emotional upsets it may well give rise to irrational decisions and distractions. A manager at the anxious end of the spectrum is less likely to be an excellent manager, and certainly not one that is at ease and enjoying the work. It might also be thought that anxiety and conscientiousness might be related, but they are in fact two separate and independent traits, so it is possible to have an unconscientious person who is very anxious in that they might fail to carry out a job on time and then spend the rest of the week worrying about the consequences.

Anxious people are also likely to avoid risky situations, though this approach is also related to the openness trait. Most managers will tend to be what are called ‘risk averters’ in contrast to ‘risk preferers’. Risk averters will choose options, where there is a choice, that tend to have a more clearly definable and known outcome. For example, if a fixed-price contract is available for pre-selling a product yet to be completed, in contrast to waiting to see what the open market brings, they will tend to take out a contract. This fixes the price they will receive. Of course, the price offered relative to the estimated open market price will also be a factor with a decision, depending also on the degree of risk aversion exhibited by the manager. On the other hand, a risk preferer may well choose the open market situation, though again this will depend on the relative price forecast and the degree of risk preference. It is relatively easy to find out a manager’s risk attitude through tests which are referred to below.

While the five-factor model of personality is generally accepted as the most useful in modern psychology, other models do exist. For example, the Myers–Briggs model (Keirsey, 1998) is popular and readily available. This model assumes people can be divided into 16 types based on the factors:

- extroverted (E)–introverted (I);
- sensory (S)–intuitive (N);
- thinking (T)–feeling (F); and
- judging (J)–perceiving (P).

Each pair expresses the extremes in the factors, so people are classified according to which end of each they fall. The four types then give rise to 16 possible combinations, so a person might be, for example, ENTJ in which case they are extroverted, intuitive, thinking and judging. A simple test enables classification.

Examples of the questions in the Myers–Briggs test include: ‘Does interacting with strangers energize you, OR tax your reserves?’, ‘Are you more comfortable after a decision, OR before a decision?’ and ‘Common sense is usually reliable, OR frequently questionable?’ In total there are 70 of these questions. The answers tell which of the extremes of each of the four factors the person is. A template on which the answers are recorded is used to add the scores and provide a classification of the person’s type. Some of the earlier work relating personality to business management used this test (e.g. Young and Walters, 2002; Furnham and Crump, 2015).

Still, other models have been proposed. An example is the frequently used 16-factor Personality Factor (PF) model, which does have strong correlations with the five-factor model (Cattell *et al.*, 1970). Examples of the traits are:

- outgoing–reserved;
- unemotional–emotional;
- cheerful–sober;
- tense–relaxed;
- conscientious–expedient;
- radical–conservative; and so on.

The similarity with the five-factor model is clear. Contemporary research suggests that many of the 16 factors are highly correlated with one of the five factors so that in a search for parsimony and the core traits, the conclusion has been that the five-factor model is more robust and useful.

Personality and Tests

Determining a person’s personality can be approached in two ways. Careful observation can give many clues, especially if the observation can continue over a year or so during a professional relationship, and the observer knows what to look for. Anxiety, for example, tends to become obvious through discussions over the selection of alternatives, and through general discussion over outcomes. Similar comments apply for assessing a manager’s risk attitude and the other personality traits.

The other alternative that enables assessing the degree of the various traits is to use a formal question-and-answer test, though such pencil-and-paper tests can be misleading when they are not worded well and the manager finds it difficult to express their real feelings and attitudes. In the end, use of a formal test as well as astute observation will provide the best of both worlds. Formal tests do not provide absolute measures, for there is no absolute benchmark like there is for things such as distance (the metre standard held in Paris, for example, though to be readily available the definition is now based on wavelengths of light from an isotope). However, data from a wide range of managers provides relative benchmarks.

As noted, there are many personality tests based on the various theories of personality. For the five-factor model, the common test is referred to as the NEO-Personality Inventory – Revised (NEO-PI-R) (Costa and McCrae, 1992). This test has 240 questions, 48 for each of the five factors. To administer the test you must normally be a registered psychologist before the material will be licensed to you. Given the length of the test, and the copyrighting associated with it, including the payment required for licensing, as well as the fact that it is designed for general use in contrast to being specific to managers and the terms they specifically relate to, it is not very practical for manager use. Accordingly, a shorter and manager-directed test has been developed and tested. This is described below. It must be noted that there is a shorter version of the NEO called the NEO Five-Factor Inventory (NEO-FFI) made up of 60 questions with 12 for each factor.

With a shorter test, some of the nuances of personality are lost, but when considering managerial ability, these nuances are not so important. The five-factor model has what are called 'facets' for each base trait which the long tests measure. The facets include:

- Openness: fantasy, aesthetics, feelings, actions, ideas, values.
- Conscientiousness: competence, order, dutifulness, achievement striving, self-discipline, deliberation.
- Extroversion: warmth, gregariousness, assertiveness, activity, excitement seeking, positive emotions.
- Agreeableness: trust, straightforwardness, altruism, compliance, modesty, tendermindedness.
- Neuroticism: anxiety, angry hostility, depression, self-consciousness, impulsiveness, vulnerability.

It will be noted that some of the facets are unlikely to be strongly related to managerial ability. Examples include modesty, aesthetics, altruism, tendermindedness and excitement seeking.

It was considered that a manager would have little difficulty in answering 25 questions (actually statements) given that, in any test situation, there would also be other tests included. Thus, a test of five questions for each basic trait was constructed and given to several samples of managers for testing.

Given that the objective was to assess a manager's personality with respect to management, the test was referred to as a management-style test. As noted, this contains 25 statements each of which the manager is asked to rate for the degree of truth on a 1 to 5 scale ('true' = 1 ... 'untrue' = 5). The questions are presented in Appendix 2A to this chapter.

Given that the data was collected from many hundreds of managers, the correlations between the questions were assessed to determine the underlying factors. The results clearly indicated there were six management-oriented traits that reflected a manager's management style (Nuthall, 2006).

Based on the questions making up each trait, they were called:

Trait	Questions making up the trait (no. in the question set)	Average score
Correctness anxiety	5, 10, 13, 20	12.8
Conscientious planning	8, 9, 16, 17, 19, 21, 24, 25	20.4
Thoughtful creativity	2, 9, 11, 12, 15, 16	14.3
Community spirit	7, 15, 18, 23	11.2
Consultative logician	1, 2, 3, 4	9.9
Benign acceptance	6, 7, 14, 22	11.8

Beside each trait are listed the numbers of the questions that make up the factor (see Appendix 2A) and the average score for the sample of managers across all business types. The score is obtained by adding the statement 'truth score' (1 to 5). The averages vary due to both the tendency to be at one end of the scale and the number of questions making up each trait. It will be noted that managers, on average, do not tend towards exhibiting 'correctness anxiety' (maximum possible score is 20 for completely 'not true'), are middle of the road for 'conscientious planning', definitely tend towards 'thoughtful creativity', exhibit a slight tendency away from 'community spirit', but lean towards being 'consultative logicians', while being slightly the opposite to 'benign acceptance'. These are, of course, the averages for a particular population of managers.

For good management it is better to have a low score in traits 2, 3 and 5, and probably also in trait 4, whereas a high score in traits 1 and 6 may create problems.

It will be noted that some of the questions appear in more than one trait (question 16, for example). This is because some of the elements of what is embodied in a question impact on more than one trait. Strictly it would be better to test more questions to ensure complete independence. Each question has a weighting on how much it contributes to a trait based on a statistical technique known as 'factor analysis'. For example, the weightings of the questions in trait 2 are 0.39, 0.43, 0.35, 0.56, 0.49, 0.38, 0.63 and 0.62, respectively. When naming a trait, it is necessary to take these weightings into account. Trait 2 has questions 17, 24 and 25 as being particularly important.

With respect to the five-factor model of personality, it is clear that 'correctness anxiety' is similar to neuroticism (anxiety), 'conscientious planning' to conscientiousness, 'thoughtful creativity' to openness, 'community spirit' to extroversion, 'benign acceptance' to agreeableness and 'consultative logician' relates to extroversion as well as conscientiousness and openness. This amalgam is clearly important.

Every manager will have a unique score in the style test indicating, assuming they ranked the statements truthfully, their particular phenotype. This will impact on their managerial success in conjunction with the other important factors. It is important to understand this interaction between style and success when assessing a manager as it helps explain the outcomes being achieved. It is common to assume relatively poor outcomes are due to inappropriate product and process selection. While this is probably true, improving the situation is often not just a matter of deciding what the selections should be, but of improving the skills of the manager so s/he is better able to make the selections her/himself. Moving in this direction starts with testing the manager to define her/his management style. Astute observation of a manager as s/he operates adds to this assessment, though in some cases this will be all that is available. Understanding the human condition will greatly enhance the success of the observations.

Manager Hank, one of the case studies, has an interesting managerial style. In line with his risk-averse attitude (see below) he tends towards having a moderate-to-high level of 'correctness anxiety' (score of 8 out of 20, with the lower the score the stronger the factor – the population average is 12.8) and this leads into his need for 'conscientious planning' (21 out of 40) being one of the more important style factors. Hank is constantly planning and replanning his activities with 'paper' experiments using a range of software packages. Hank tends to concentrate on the business as a focus for his energy and is strong in 'thoughtful creativity' (10 out of 30) with many ideas bounding from his fertile mind. He also is keen on mulling over ideas with others, with his 'consultative logician' (5 out of 20) score being particularly strong. Hank is also a tolerant person, and this is expressed in a similarly strong score in 'benign acceptance' (5 out of 20). Hank reckons these figures are a true reflection of his style, and the fact that he comments on his own personality suggests he is introspective and thoughtful about his own capabilities. This self-realization is an important factor in ensuring the true situation is honestly recognized.

It is interesting that Margrave, another very successful manager, has a similar managerial style for only some of the factors. Margrave's 'correctness anxiety' score is much higher at 17 out of 20, relative to the population average of 12.8. This means Margrave tends not to worry about issues, and this complies with his risk attitude as determined by the risk test given below. Margrave has a risk-neutral attitude, meaning he is prepared to take the good with the bad and seldom makes decisions based solely on their riskiness, or lack of it. Margrave is slightly less of a 'conscientious planner' (score 23 out of 40) than Hank, and this shows in the lack of detailed records he keeps: 'Why keep records which I can't see a use for?'

Hank might argue that you do not always know what records and information you might want in the future. When it comes to his 'thoughtful creativity', Margrave is right on the population average of 14 out of 30. A casual observer would have said that Margrave was very thoughtful and creative as shown by the wide range of activities he has instituted, so perhaps the book test was not a successful representation of his true style. Where observations question test results, further investigation is warranted.

Margrave has a strong 'community spirit', with the low score of 7 out of 20. This is the same as Hank and reflects that both managers are involved in community activities. Margrave, for example, is involved in manager committees that review the state of business and make policy suggestions to the administrators. Margrave is less of a 'consultative logician' than the average with a score of 11 out of 20 (population average 9.9) but has an accepting management style with a 6 out of 20 score for 'benign acceptance'. Observation backs up these formal scores.

Attitude to Risky Situations

Related to managerial style is a manager's attitude to risky situations. A risky situation is one where outcomes cannot be predicted with certainty. Thus, for example, a crop yield can take on many values depending, primarily, on the weather and disease impacts. While the range is probably predictable, all that can be said about the yield and profit is that various levels are possible, each with a degree of probability. Many such examples abound (a factory might have variable output due to skilled labour availability and also receive a variable per product price on a volatile market, and so on). The more variability, the more risk. This risk can be reduced through various management techniques; an obvious one in the case of crop yields example is the use of irrigation, though this may be profitable in its own right. Whatever the case, irrigation will certainly reduce the degree of risk. Managers that invest in irrigation just to reduce the risk are referred to as risk averters (Fig. 2.2). In contrast, a manager who enjoys taking risks is called a risk preferer, though there will be all shades between the extremes. A risk preferer, for example, might well not bother with sickness insurance as the possible benefits from not paying the premium more than compensate the downsides of sickness-created loss of work time.

Fig. 2.2. Risk attitude influences decisions. In this case storing more than farmer clients' average yearly demand for winter feed supply has been important.

A manager's attitude is undoubtedly a function of past experiences, their personality and the situation they find themself in. If, for example, the mortgage repayments are high, a manager may make conservative decisions to ensure the payments can always be met. The outcome of a poor year in which the mortgage is called in is just too horrendous to contemplate even where the manager's basic attitude may tend towards risk preference.

There is nothing good or bad about a particular attitude. What is important is that a manager is able to take note of his attitude and act accordingly. If his decisions do not match his risk attitude, it is likely that he will be concerned about his situation, which in turn will lead to a reduction in his general level of satisfaction.

This is not to say a manager should not examine his attitude and, perhaps, make efforts to change it. An overanxious manager is likely to be a severe risk-averter, leading to poor decisions. Making an effort to be less of an averter, perhaps with the help of a spouse and other friends, might well have a considerable pay-off in terms of achievement and satisfaction. The same comments apply to extreme risk-preferers who may well gamble away assets inappropriately.

There are a number of tests to assess a manager's attitude (e.g. Anderson *et al.*, 1977; Charness *et al.*, 2020). However, observation of everyday actions may well give a good indication of a manager's attitude. An informal test might well be constructed to suit the particular environment. An example of this kind of test is given below, in which a number of typical decision situations is offered with a range of answers each of which reflects aversion or preference. Taken together the answers lead to a conclusion.

Question set to assess a manager's risk attitude

The subjects of some of the questions will need changing to suit each particular type of business (domain).

1. **Would you prefer to**
   - **i.** take out a fixed-price contract on your output of US$3.80/kg? OR
   - **ii.** accept what the market offers at the time?

   You anticipate the price will be US$3.90, but there is a good chance (40%) that it will be US$4.20, but it could be as low as US$3.20 with a 20% chance.

2. **If you have irrigation, do you have**
   - **i.** an irrigation system that does NOT get used to its full capacity? OR
   - **ii.** an irrigation system that IS used to its full capacity?

3. **Insurance records show that** the chance of your main storage building and its contents going up in smoke from accidental causes is only 0.1% (i.e. one chance in 1000 – would expect the building to burn down once every 1000 years). The building stores 1000 output unit bales (currently selling at US$5 each).

   If they want to charge an annual premium of US$1180,

   WILL YOU,
   - **i.** insure the building and contents? OR
   - **ii.** NOT insure the building and its contents?

   The replacement cost of the building is US$4000. You normally have it full.

4. **A wholesale buyer is offering you US$3.60/kg** for one of your products, but you have a suspicion the market is going to lift. Your reading of the world scene is that the price could be as high as US$4.10 at the next sale, though as low as US$3.40 is certainly not out of the question as the world scene is somewhat shaky. In mulling over the situation you come to the conclusion that there is a 50% chance of getting close to US$4.10, a 25% chance of US$3.75 and a 25% chance of US$3.40.

   WILL YOU
   - **i.** sell to the buyer? OR
   - **ii.** sell at the next auction?

5. **You have been contemplating increasing** the size and number of grading machines in your packhouse, as in some years you seem to have a greater supply of raw product than the current machines can handle. The trouble is, given a series of average to bad seasons, you would struggle to keep the increased investment fully

operating. Your calculations and hunches suggest the following:

a. If you stay as you are – average profit per unit of capacity will be US$520.
b. If you increase the capacity 10%, profit in a typical year will be US$450/capacity unit, profit in a good year will be US$550/unit, while profit in a poor year will be US$380/unit.

Chance of a good year 40%, poor year 20%, typical year 40%.

WILL YOU

i. stay as you are? OR
ii. increase capacity by 10%?

6. Which best describes your betting actions?

i. seldom take LOTTERY tickets and spend less than US$1000/year on sports betting, OR
ii. do take LOTTERY tickets more than occasionally and spend quite a lot on betting.

7. If you rely on cash reserves to cover the low sales over the winter period, do you

i. regularly have more than 30% of your reserves left over each spring? OR
ii. seldom have much left over?

8. If you have (had) a mortgage, do (did) you use

i. the floating-rate option if available to you? OR
ii. the fixed-rate option?

Skip this question if the manager has never faced this choice.

9. For your extensive subdivision plan (perhaps imaginary) requiring fencing, improving the water supply and some road making, would you accept

i. a fixed-rate 10-year mortgage at the rate of 7.5%? OR
ii. a floating-rate mortgage?

Your discussions with bank managers, and your reading, suggest the following: Worldwide the long-term average rate is likely to slowly decline, and it is almost certain that the average exchange rate will stay much the same as it is now, though there will be minor variations. However, history tells you that there is little that is certain about interest rates and you reckon the average rate for floating-rate mortgages could go to 8.5% with a 40% chance, but equally there is a 40% chance it could actually decline with the average over the 10 years turning out to be 6.5%. Will you take a FIXED- or FLOATING-rate mortgage?

Answers

The number of times option (i) is selected relative to the number of times option (ii) is selected will tell you the person's risk attitude. If (i) is greater, the person is a risk averter, and vice versa. The degree of aversion or preference depends on the balance.

For question one, the second choice has the highest pay-off on average, but it is riskier in that it may pay less than the first choice. For question two, people not using their irrigation to full capacity, assuming this is possible, represents risk aversion in that the irrigation is being used for insurance rather than straight profit. For question three, the least-cost option is to take the risk yourself, but few would do this. Thus, buyers of insurance in this case represent risk aversion. In the case of the selling policy in question four, option (ii) is the most profitable, but is clearly riskier in that the price could be both lower and higher. A risk averter will, therefore, choose option (i). In question five, staying at the current number of grading units is both more profitable and safer. However, if someone was to choose to increase capacity, they are very much gambling on the chance of a very good year, and are, therefore, an extreme risk-preferer.

It is clear that choosers of option (i) in question six will be risk averters as lotteries are always stacked in the favour of the organizers. In the case of the cash reserves (question seven), a risk averter will always select option (i) due to its 'safeness'. In question eight, the safe and sure way (risk averter) is to select a fixed interest rate, thus giving surety; and finally, in question nine, the choices are equal in an average financial sense, but someone choosing option (i) will be looking for surety and exhibit risk-aversion tendencies.

It will be noted there are nine questions, so one of the totals for each of the options will at least dominate the other by one. The degree of domination gives the strength of the attitude.

It will be clear from these questions that, in a practical sense, careful observation of how a manager operates will provide a quick indication of risk attitude. A consultant must weigh up whether trying to change the manager's attitude will be worthwhile, leading to greater satisfaction. In some cases, attempting change might cause more problems in that the person might become quite anxious over the whole affair. Caution is required.

The case managers found the questions easy to answer. Discussion and observation of their systems indicated that the test results were in line with what you would expect. Hank chose option (i) six times and option (ii) thrice, indicating that he is a relatively strong risk-averter, and this shows in, for example, his constant use of irrigation and reasonable reserves. Margrave, on the other hand, chose both options (i) and (ii) four times, with one question not answered as it was not applicable (the irrigation question). Margrave has a clear risk-neutral approach to management and, as a consequence, is content to stretch the business's resources, but he also plans on selling off current assets when the need arises in a bad year.

Despite his risk-averse attitude, and his relatively high 'correctness anxiety', Hank does not stay awake worrying about the situation if he has carefully thought through the problems and is comfortable with his conclusions. Perhaps his management style is the reason he puts so much time into researching problems and doing all the sums. Having finished his analysis, he 'rests easy', knowing he has fully investigated the situation and has come to the right conclusion. Hank does, however, recognize that as a specialist manager 'all his eggs are in one basket', meaning a downturn in the world market is not compensated through diversification. Hank commented that his children have been involved in the wine industry, and he wondered about establishing a vineyard nearby. The initial research, however, suggested the number of degree days over the summer would probably preclude grapes having a sufficient brix reading to make good wine. Perhaps this is something for the future, once cool-climate viticulture has been further explored.

Intelligence and Its Links to Management

Few would doubt the importance of intelligence in successful management. Intelligence has been studied by many over a long period of time, yet there are still many opinions on exactly what is meant by intelligence. Early definitions specify intelligence as the 'capacity to learn from experience and the ability to adapt to the surrounding environment' (see Sternberg, 1995). Generally speaking, most popular belief on intelligence regards it as the ability to understand situations and ideas as well as sort out problems, and also involving an extensive store of information that is readily acquired. That is, being logical and having a good memory. Clearly these two components are critical to good management. Personality, on the other hand, would commonly be regarded as the emotional and people aspect to being human.

In that there are many components to human lives, intelligence has many aspects. Gardner (1993) lists the components as:

- Verbal–linguistic: i.e. the ability to use words and language.
- Logical–mathematical: i.e. the capacity for inductive and deductive thinking and reasoning, as well as the use of numbers and the recognition of abstract patterns.
- Visual–spatial: i.e. the ability to visualize objects and spatial dimensions, and internal images and pictures.
- Bodily–kinaesthetic: i.e. the wisdom of the body and the ability to control physical motion.
- Musical–rhythmic: i.e. the ability to recognize tonal patterns and sounds, as well as a sensitivity to rhythms and beats.
- Interpersonal: i.e. the capacity for person-to-person communications and relationships.
- Intrapersonal: i.e. the spiritual, inner states of being, self-reflection and awareness.

Clearly some of these so-called independent intelligence traits do not have a bearing on management. Of course, other psychologists would not totally agree on this list.

In a general sense, intelligence is often divided into two:

- fluid; and
- crystallized.

Fluid intelligence covers broad basic reasoning, which is largely genetically sourced, whereas crystallized is fluid intelligence as expressed in a particular culture. Adjectives that have been used to define the components of fluid intelligence are:

- inference;
- induction;
- memory span;
- intellectual speed;
- visualization; and
- retrieval capacity.

For crystallized intelligence, words used include:

- verbal;
- mechanical;
- numerical; and
- social skills.

If anything, there is more consensus over what is called the Triarchic Theory of Human Intelligence than most concepts. This was proposed by Sternberg (1995) and has three basic components described as the cornerstones of a triangle. These are a person's:

- analytical abilities;
- creative abilities; and
- practical abilities.

Adjectives that come to mind for each apex of the triangle are:

- analyse, compare and evaluate;
- create, invent and design; and
- apply, use and utilize.

In a management sense, the three basic components are all relevant, as are the facets of each attribute. Successfully analysing alternative causes of action is a critical skill in management, as is creativity in that it is critical that a manager can create solutions to problems in the widest sense of the word 'problem' (a problem might well be the question of what the best set of products is to produce next year, for example. And then there are the literal problems, such as what to do if one of the manufacturing resources is running out.) In addition, of course, practical knowledge and ability is absolutely necessary in many SME rural businesses in that success depends on physically getting the business and its productive resources into an appropriate state. While a manager might employ people to carry out the physical tasks, it is not possible to plan, organize and control outcomes unless a good understanding of the production process exists.

It has been noted that experience of the right kind is probably an important aspect of managerial ability, as it is in many occupations. Learning from experience is not just a matter of existing through various situations, but a matter of:

- observing the situation;
- analysing it; and
- learning the lessons available for future reference.

Thus, a suitable intelligence succeeds in obtaining the most from experience, and of making use of the stored lessons in solving future problems. As noted earlier, the making of an expert involves learning from experience.

This brings up the memory aspects of intelligence. Memory is generally regarded as limitless, though you may not agree when it comes to recalling the things you have learnt. It is a matter of training yourself to remember and recall items at the correct time. Research, and common observation, suggests that memory is divided into short- and long-term components. You constantly observe sights, sounds, smells and feelings, and these move into short-term memory where the inputted observations are processed for meaning and relevance. Much of the data is immediately discarded. Think back to 5 min ago – do you still remember the sounds and objects you observed? Much of the detail has been lost, or at least not recallable. If, on the other hand, you heard, for example, a violent shutting of a door you probably still have this in mind. The reason is that the short-term memory and the processing system noted this auditory input as being unusual, and so it was passed into long-term memory for later recall. Essentially, the processing filters material to assess whether it is valuable, and if so pushes it through to long-term memory where it can be stored, perhaps forever, and recalled when required.

People who have suffered physical brain injuries usually comment that they cannot remember the knock and what went immediately before it. This happens as the blow interrupts the processing stage and so the conclusion is not sent through to long-term memory. The lesson from this situation is that a good manager will train to ensure the appropriate processing occurs with the useful material being shunted through to long-term memory. Good training will ensure that the crystallized intelligence is well honed for the management task.

Short-term memory, and the processing of material entering, are thought to be limited. However, this probably varies with a person's level of intelligence in that we all know that some people are better at observation, and sorting out what is important and the relationships between the observed information. You will be aware that telephone numbers are usually limited to seven digits. Tests suggest seven is the maximum number of items fitting into immediate short-term memory (five to nine items seems to be the range of the limits). If too much information arrives too quickly, a considerable quantity is lost unless the person has very high processing speed (thus speed is considered part of high intelligence). In some domains, for example, primary production and processing, the speed skill is probably not as critical as in other industries in that action does not usually have to occur instantly, so it is possible to observe and consider before action in a day or two.

The skill of committing material to long-term memory usually involves rehearsal, thus giving a strong imprint. The more rehearsal, the better. With practice and a knowledge of how to rehearse long-term memory, storage and retrieval can be enhanced considerably. In effect, intelligence can be improved, particularly the crystallized aspects.

It has been mentioned that intuition, or tacit knowledge as it is sometimes called, is important. In an intelligence sense, as was noted for experts, high intelligence assists this process of taking into long-term memory a store of helpful tips that creates a successful intuition. Kelly (1992) talked about the concept of holding decision rules through what he called 'constructs'. The totality of a person's constructs makes up what can be called 'intuition'. Kelly believed in 'man as a scientist' in that 'man' was always striving to sort out rules by which he could operate. Thus observation led to the idea that people invoke a particular construct that suits the situation – that is, a decision on what action to follow. With experience, the constructs became better informed and appropriate to each individual. If the person is uncomfortable with a construct, searching occurs until the decision maker finds one acceptable to her or his psyche. Such constructs might relate to aspects of a person's personal life, or in this case, their professional life. When you analyse how you make decisions, it is seldom a full and clear analysis in contrast to a fairly rapidly 'arrived at' conclusion. You have brought out a construct. Clearly, a manager's intelligence level will impact on the process of creating suitable constructs (sometimes called 'rules of thumb', or heuristics).

What parts of intelligence are important to management? Logic would suggest that a good memory of retrievable material of relevance to the business domain of interest is a starting point. Also important is the ability to learn from experience. It is not possible to appreciate the important aspects of the production processes and functions in the relevant domain, appropriate marketing, labour selection and management as well as financial organization without actually being involved. The term used in days gone by was 'the university of hard knocks'. And it is true that no amount of book knowledge will enable a manager to determine the important aspects of successful management. Consequently, the manager who can learn successful 'constructs' from practical observation and experience will more likely succeed than others who seldom learn from mistakes and experiences.

The third component is creativity. Problems frequently require innovative solutions, and certainly opportunities need observing and exploited using, possibly, new production processes that suit the particular business. Creativity is related to imagination and both these attributes are important, provided the results of the creativity can be turned into practical solutions. Practicality is an important attribute of intelligence where production is biologically based. There is nothing mechanical nor predestined about it, so textbook solutions are seldom appropriate in a rote sense.

Moreover, an ability to understand the economic principles and decision rules relevant to production and processing is critical, as is a knowledge of how to calculate the value of the objectives as a result of some proposed decision and change. Generally this attribute might be classed as skill in business-based calculations. This does not mean being a maths guru, but knowing what items to include in a simple sum, and how to summarize the result. For example, given an average condition decline rate for products, and accordingly the number needed to be removed to maintain quality, and the number that need to be abandoned due to poor production characteristics, how many must be produced each year in order to, say, increase the market share and sales by 5%?

And another example might relate to the profitability of devoting a fixed proportion of the resources to a new enterprise such as a new florist shop. What is the lost net income from the resources diverted in relation to the increase in income from the new enterprise? This involves many simple estimates and calculations involving, for example, the labour freed up, the labour required, the timing of all these events (labour cannot be stored), and so on for all the other resources such as the machinery and working capital. Generally, this whole area of economic principles and simple business sums is part of general logic skills, as are other components of good management.

Finally, given the physical nature of many business types, an ability to envisage shapes is another aspect of intelligence that is desirable. For example, in a design and building business, the plan might be to create more sub-buildings by subdividing existing areas. In planning this it is important the manager can work from a map to estimate earthworks, fence lines and building structure material to enable organizing contracts and pre-ordering of the appropriate materials. Another example is an ability to assess the quality of created structures to enable assessing their condition and decide whether further modifications are required for structural strength. A further example involves pastures. With practice it is possible to judge the dry matter of a field from simple eye observation so that this judgement, together with a knowledge of the species and their stage of development, enables an assessment of the energy and protein available. Again, this ability to assess the physical observations and their significance is a valuable attribute in a production system where it is not possible to be constantly reverting to expensive meters and expert consultants.

In summary, the important attributes that relate to intelligence are:

- memory;
- an ability to constantly improve the store of successful 'constructs' created through experience;
- creativity (imagination);
- calculational ability in its widest sense; and
- an ability to work with shapes.

General logic is also important. Working out just how good a particular manager is with respect to each attribute can come from observation as the manager carries out day-to-day activities, or through a formal test, or a combination.

Intelligence and Tests

There is a wide range of intelligence tests available. Generally, most managers are somewhat dubious about being asked to sit for such tests. Furthermore, there is no general data linking the results of general tests to managerial skill, though you would expect this to be the case. There is, however, data that links intelligence tests to school results, and also to tertiary study. Consequently, educational results can give a good idea of general intelligence as defined for the purpose of the tests. Similarly, astute observation of a manager when working closely with them on, say, setting up plans and courses of action, will give a good idea of what might be called 'business intelligence'. And then, as outcomes are observed, and similarly, the thinking process a manager uses, such estimates become refined.

Alternatively, or perhaps in conjunction with, tests designed for the intelligence attributes important to the domain of interest might be created and used. Such tests need to be set up for the specific environment and production type relevant to the managers of interest and generally relate to the particular business type, though some questions can be universal.

For example, it is no use asking a glasshouse manager questions that relate to road strength calculations as they will be unfamiliar with the relationships involved. Questions on simple logic, however, can be universal.

The abilities that cannot be tested are practical skills and understanding, so judgements in this area must rely on observations of past outcomes and current activities.

Listed in Appendix 2B is an example of a test used for broad acre sheep business. It will be clear that many of the questions would need changing for, say, food factory managers. The test is referred to as a 'managerial aptitude test' in contrast to intelligence, as it is this skill which is being assessed.

To score the test, the number of correct answers is added up. For some questions the answer is simply right or wrong, but for others a degree of truth exists. In this case, a suitable scoring system template is necessary. Things like creativity are very hard to test for, and in this case scores were based on the number of realistic ideas produced. Scoring the lessons from experience is also difficult, so, similarly, the number of lessons that were realistic provided the score. For calculations, an answer within a reasonable range of the true answer was accepted as correct. It will also be noted that the section labelled general was largely used to assess simple logic.

This test was given to a large number of managers of all types, with results from 490 being included. The scoring was normalized to provide an average score of 100 in the same way as standard intelligence tests. The distribution of results is provided in Fig. 2.3.

The average of 100 will be noted, and the distribution is similar to normality as shown by the dark line. It was found that the score was significantly correlated with age, education level (years of schooling or tertiary study) and the average percentage score attained in the last year of formal study. The aptitude quotient (AQ) was predicted by the equation:

$$AQ = 87.5 + 7.0E + 0.1P - 0.5F$$

where E is the education level based on a score of 1–5, where 1 = 'primary schooling only' through to 5 = 'two or more years of tertiary education'; P is the average percentage grade for their last year of formal education; and F is a code reflecting the type of business (complexity). The equation and individual variables were all significant.

A good question is whether using these kinds of tests gives a better idea of aptitude than simply using knowledge of the manager's education background. The answer is mixed. The predicting power of the equation was 27%. Thus, 27% of the variability is explained by the equation, leaving 73% due to other reasons. Thus, while the education background is useful in assessing a manager's ability, many other factors, perhaps experience, are also important. This is borne out by the fact that some very good managers do not have much formal education. Also, it will be noted that the number of years of education is rather more important than the final-year grade.

Knowing a manager's aptitude is important when considering how managerial ability might be improved. The aptitude score clearly indicates the possibilities with respect to the potential to manage complex operations. However, the manager is a total package and while aptitude might be limiting, this may well be compensated for in part by, say, a very suitable personality. Thus, when looking at what might be a suitable strategic plan, both aspects must be taken into account.

Furthermore, when providing assistance to a manager, a knowledge of the manager's skills within all the important aspects of intelligence indicates those areas where most help is needed. For example, perhaps the manager is not particularly good at calculations, therefore help in sorting these aspects will be beneficial.

While the genetic side of aptitude is fixed, from the description of intelligence it is clear improvement is possible. That is what the education system is all about, but in this case we are dealing with people in the workforce, so changes come about either by organized educational programmes or self-education. In most countries, distance learning is available so given a strong motivation there are opportunities to tackle improvement. An extension group, or consultants, can similarly organize suitable programmes, perhaps on a one-to-one basis in some cases. Where a manager seeks help from a professional, they can consider taking extra time to explain and train in contrast to simply providing a fixed plan or action.

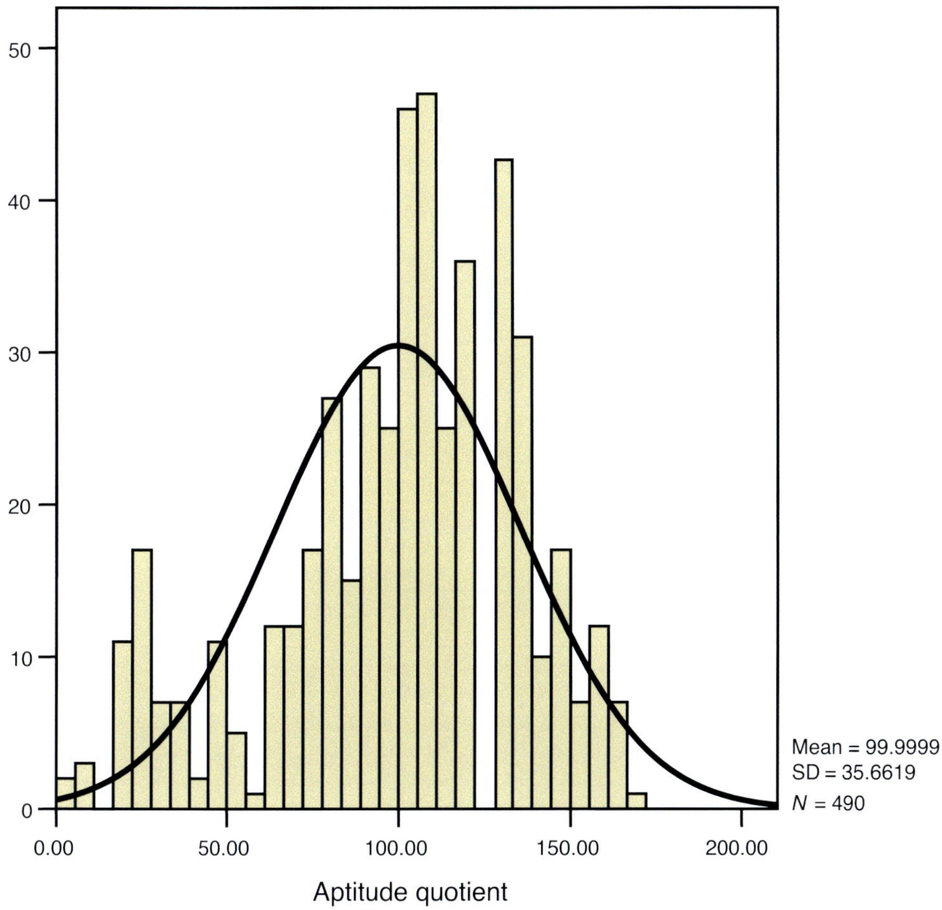

Fig. 2.3. Aptitude quotient distribution frequency.

When given the aptitude test, the case managers did find answering was a challenge, but given the situation they were determined to take a reasonable time for fear of being thought of as unintelligent. They need not have worried. One obtained a score of 154% and the other 174%. These managers are clearly well into the top section of the distribution. Both have some tertiary education so perhaps this is one of the major contributors, though their background and personality will also influence the motivation to take and succeed in tertiary establishments. Certainly Margrave maintained that while he had attended very many short courses, where he gained most was from the tertiary education which forced him to develop his problem-solving and analytical skills. He also mentioned the importance of being generally critical in a positive sense, and that this skill was very much a product of his tertiary education. As an aside, it is interesting to note that the survey work referred to in creating the aptitude distribution, and also other survey work, indicates the amount of tertiary education found among managers is much more than that in the general population (around 30%).

Motivation, Objectives and Emotional Intelligence

You would imagine a manager's motivation and objectives would influence managerial skill. A motivated manager will seek to be good at his job and seek out ways to improve. And his objectives may well influence the shape of decisions made, and therefore impact on outcomes. In assessing a manager, it is important to know their objectives as someone not primarily motivated to achieve, say, maximum profit, will operate on a different plane to someone who is.

For example, Margrave, one of the case study managers, noted he had 'unfinished business' when he left school, for while his early life was in a business, he had not been involved other than carrying out physical chores. His family were almost secretive about business and decision making. Margrave believes his interest in his industry was almost 'innate' with the passionate interest being 'in my genes'. Margrave could not wait to sign up for short courses and get to work in businesses. Located within easy distance was a training establishment that constantly offered 'live-in' short courses. Margrave enjoyed them immensely and absorbed the lessons in an almost sponge-like fashion. This real passion gave rise to an extremely high motivation, which drove him to learn and learn despite long hours and exhaustion from the constant physical work. In the end, Margrave rated himself as being 'quite good' as a manager, and noted that the business had increased in capacity by over 30% in recent years. You cannot doubt his skill and motivation.

Where managers have different objectives, can it be decided which are more efficient and managerially expert? Such an assessment, of course, depends on the objectives – so a manager who wishes to, say, operate an organic and low greenhouse gas emission system can be just as managerially efficient as a manager maximizing profit. An observer of management must therefore take the 'business's' objectives into account, as well as the motivation to succeed. In some cases, it will be possible to observe motivation, and consider whether inappropriate motivation is getting in the way of efficiency. Changing a manager's level of motivation, on the other hand, might be more difficult.

The remainder of this section contains a discussion on motivation, its relationship to personality (managerial style), and considers how objectives might be quantified remembering that many managers are owner/operators and so determining objectives is easier than if the ownership is complex and divorced from the business. Also discussed is the psychometric test for emotional intelligence (EI) as some believe what it embodies relates to managerial outcomes. As the name suggests, it is to do with emotions related to getting 'things' done.

There are many theories on motivation: what it is, and how it might be measured. Furthermore, the jury is still out on whether it is a separate human trait, or whether it is an expression of personality and a sub-component reflecting the basic traits. However, it is an obvious trait to discuss when considering managerial skill and, therefore, must be included here.

Most people would consider 'motivation' to involve not only an ability to think of, and initiate, a project, but also the control of the direction of its conduct and final completion, including persistence to finish off what is started, if this is indeed rational. Further, motivation probably relates to a 'need, want or interest' a person has that propels them in a certain direction. Frequently the motivation to achieve a goal will stem from 'intrinsic motivation' meaning that the drive comes from the pleasure it brings, or from a feeling that the task is important, or that it is morally required. All these drives are internal, with success leading to the person feeling better about themselves. In contrast, the drive to achieve might be 'extrinsic' meaning it comes from outside the person. An example would be the bank manager demanding repayment of an overdraft within a defined period.

Self-control is an important part of motivation. This refers to the ability to perform the tasks required to achieve the goal or need in a timely manner. Many people have the right aims but actually carrying out the tasks on time is another matter and requires self-control. Sometimes the degree of self-control depends on the needs. We all need to eat, but few in the Western world find it difficult to raise the self-control needed to achieve basic bodily needs; indeed the control required is in not eating too much. As the 'needs' become more divorced from these basic needs of life and society, it often becomes more difficult to carry out what is required. Thus, the motivation to become a perfect manager involves more and more self-control and dedication.

There are theories on what drives people to achieve their goals. What drives you to study? Perhaps it is your conscience that will not be content until you have achieved a goal?

You have an internal drive that forces you to continue until your system reaches a comfortable equilibrium. Or perhaps you have some external incentive to drive you on to success. An example might be a promise of a substantial reward once you have achieved success. Clearly your emotions are involved in the drives. Emotions give rise to a state somewhere on the continuum of unpleasantness–pleasantness, so if you are feeling uncomfortable about a situation you may well strive to fix the problem and move yourself into a state of pleasure. Just how much effort goes into creating this change will depend on the degree of unpleasantness and your means to change the situation. If you have no control over the situation, you may change your feeling as you rationalize that it is impossible to do anything about the situation. Some people, however, while realizing it is not possible to alter the situation, still find it unpleasant and consequently lose sleep. You might class such people as 'anxious'.

Emotions can also cause physiological responses such as a rise in your heart rate and associated panic reactions. On the positive side, feelings of achievement might cause relaxation and a balanced bodily state that even perhaps leads to good health. In a business sense, such emotional reactions can stimulate appropriate action like getting out of bed in the middle of the night to attend to a problem such as potential flooding of the factory.

One wonders how there can be such a variation in motivation within society. It has been proposed that there are major differences between individuals with respect to their need for:

- achievement;
- power; and
- affiliation.

You might also add to this the need for praise and positive feedback. It seems factors that encourage motivation are:

- autonomy;
- competence feedback; and
- relatedness.

It is likely that the time between action and outcome is also a relevant factor in that if action and response occur close together and it is an easily observed outcome, the motivation to achieve and complete a job is higher. We are all aware of this, particularly when studying for a qualification that might be years away.

In a business sense you can observe people who are very proud of their achievements, and others who just simply 'get on with the job' with little fuss or fanfare. The personal power factor is somewhat less important in that it is only the large business situations where considerable labour is employed that give the opportunity to feel powerful, though in some cases the 'power' is felt through being able to control a large business with, for example, many sub businesses and products. The need for affiliation is often expressed through joining, mainly, business organizations. It is interesting to note that this need is likely to be related to the extroversion trait.

In a practical sense, expressions of motivation can be seen in how people organize themselves. Some managers make lists of prioritized jobs, and as they are achieved they get crossed off, leading to satisfaction that is clearly noted. Such lists can be reordered as conditions and situations change. This process is part of self-created positive feedback and helps maintain motivation. Similarly, a good record-keeping system helps observe where achievements have occurred, particularly with respect to physical outputs relative to inputs and, thus, efficiency changes. Monthly business accounts, if properly kept, will also lead to noting changes in profit (and cash flows), and provide feedback both in a motivational and also in an economic sense.

Of course, talking about efficiency suggests that humans have a tendency towards rationality. However, while most would say they strive for rationality, in practice straying from rational decisions often occurs. No doubt this relates to the strength of a manager's motivation relative to the goals. Sometimes emotion takes the better of rational thought, leading to erratic decision making that may depend on the feelings on the day. This is where a conflict of motives plays out, so that, for example, the profit motive loses out to the enjoyment, say, of seeing perfectly presented packed boxes of the output all laid out in rows, or perhaps the draw of a family reunion in contrast to getting the accounts in on time. Humans are complex organisms ruled by a myriad of factors as this discussion has pointed out. It is the play of all these factors that leads to an individual's motivation.

While there are many tests for personality and intelligence, few exist for assessing motivation. This means astute observation is the main tool. Over time, a clear feel for both the manager's motivation and the relevant goals will surface. Attention to detail, degree of careful planning, determination to get the jobs organized and achieved on time are all indicators, as is the concern for exploring all new and possible technologies and products. Does the business come first where conflicts for time and energy exist?

One test that has been explored is called the Motivational Trait Questionnaire (MTQ; for a reference see Kanfer and Ackerman, 2000). This test is not specifically designed for any one domain. It is based on the theory that motivation is made up of three factors:

- personal mastery;
- competitive excellence; and
- motivation related to anxiety.

Each has several facets. Personal mastery is made up from:

- the desire to learn; and
- mastery (e.g. 'I set high standards for myself and work toward achieving them').

Competitive excellence has the following facets:

- other referenced goals – this relates to whether you compare yourself to others' achievement (e.g. how do your production and sales figures per employee compare?); and
- competitiveness (e.g. 'I would rather cooperate than compete' (reversed scored)).

Finally, motivation-related anxiety is made up of:

- worry (e.g. 'Before starting an important project, I worry about the consequences of failing'); and
- emotionality, which records a person's emotions related to outcomes.

The big question is whether motivation measurement does in fact tell you anything more about a manager than the results of the management-style test, or similar personality tests. From the discussion you might conclude that the general anxiety trait and the conscientiousness trait might well record most of what there is to know about a person's motivation. Further, the extroversion trait is likely to relate to feedback information through contacts with other people. Indeed, whatever research that is available does suggest there is a strong correlation between motivation and personality such that separate motivation tests add little information. Thus the comments above about astute observation. Given personality test results and observations it should be clear what level of motivation is likely.

In the case study manager Hank's background there may be specific factors that created the high level of motivation he clearly has. This is in contrast to the idea that general management style is all there is to know about a person. In reality, motivation, in some cases, will be due both to the manager's genes and their general experiences as well as any notable and especially memorable experiences. Hank recounts how he was heavily involved in the local business club in which, as a child, you were encouraged to take part in a range of competitions such as preparing and showing your imaginary plans. Hank could not bear to lose; he notes, 'I'm competitive. A failure is a great incentive.' He also remembers missing out on achieving at football and became quite upset as, with his competitive nature, you had to win. This competitive nature has carried over to adult life and provides a strong motivation to succeed. Hank even goes as far as to note, 'I had quite a problem of pride in my boyhood. It was cause for upset, but also determination.' This is despite his parents 'not being as driven as I am', suggesting that his early experiences were the seed for his strong motivation. But, of course, his personality would have affected his feelings over failures, for others might have just brushed over the events with hardly a touch of downcast feelings.

Examples of research relating motivation to personality include Judge and Ilies (2002), who found that the basic five traits had a multiple correlation with motivation of 49%; and Kanfer and Ackerman (2000), who concluded that anxiety was highly correlated with the motivation traits, but motivation was not related to either fluid or crystallized intelligence and is, therefore, independent of intelligence. In addition, Zuckerman *et al.* (1999) found correlations as high as 80% for motivation and the big five personality traits. They concluded that anxiety and extraversion were important traits in motivation. Judge *et al.* (2014) also related aspects of motivation to personality.

In view of these results, it is believed that developing motivation tests specifically for any specific domain will not provide additional useful information about a manager over and above management-style test results combined with astute observation.

In contrast, however, there may well be value in conducting a formal assessment of a manager's objectives. The results can be combined with the observations that anyone dealing with a manager will inevitably make as they reflect on the decisions made and the objectives they portray. The next chapter (Chapter 3) contains a set of questions that can be used to quantify a manager's objectives, or indeed the objectives of other members of the family and management team.

While a manager's motivation is highly related to his personality, the objectives held are probably very dependent on the influence of the manager's family, particularly in his earlier years, as well as the influence of the community within which the manager grew up including the effect of the paradigms expressed during schooling. Just what gave rise to the final motivation level and objectives is probably complex and highly unique for each manager. What is important, however, is making a judgement on the level of motivation held by a manager, and the objectives and goals driving the manager. Any professional coach should discuss the situation with a manager with a view to assessing the appropriateness of the goals and encourage change where this is clearly what the manager and the family want. Similarly, where motivation is a problem, counselling and support may well be appropriate in the interests of improving managerial ability and efficiency. (For further information on objectives, and a set of statements that can be used to assess a manager's objectives, see Chapter 3.)

While dealing with the personal factors of motivation, objectives and goals, it is also useful to comment on what has become known as EI. This is a relatively recent concept and covers intelligence as related to the emotions in contrast to intelligence related to the aspects normally tested in IQ tests (reasoning, memory and so on). Some would say, of course, that these aspects of a person are measured through personality tests.

Emotional intelligence (Zeidner *et al.*, 2004) is made up of:

- awareness of emotions in self;
- awareness of emotions in others;
- management of emotions in self; and
- management of emotions in others.

Clearly, a manager skilled in all these aspects is likely to achieve better outcomes than those not similarly skilled, especially where the business employs a reasonably sized workforce and deals with many contractors. It is suggested that EI is related to:

- effective networking;
- conflict management;
- stress management and adaptability;
- negotiations; and also
- listening and verbal communication.

All of these attributes are regarded as desirable in good management. Psychometric tests are available for EI and follow the usual idea of a series of questions which are then scored.

The tests are not presented here for it seems that EI is highly correlated with personality. One study concludes that the Myers–Briggs personality trait 'intuition' is significantly and positively related to higher levels of EI (Higgs, 2001). Zeidner *et al.* (2004, p. 384) also conclude that 'despite the important role attributed to a wide array of emotional competencies in the work place, there is currently only a modicum of research supporting the meaningful role attributed to EI' (Szcześniak *et al.*, 2020). Given these conclusions further discussion of EI is not considered beneficial at this stage.

Learning Style

The final test to be presented relates to assessing a manager's learning style. Good managerial skill involves constant learning, both in a skill ability sense, but also ensuring a manager's technological skill base is contemporary, as should be his knowledge of markets and regulations. Managers must be constantly learning. Just how well such lessons are absorbed will depend in part on how they are presented as we all respond best to material presented in a way that suits us.

There are at least 70 or so tests that have been developed to assess a person's learning style, and each one has an associated theory. It does seem that few of these have any scientific backing so there is little general agreement on the learning process and how it might be assessed. However, about 12 of these tests do get used. One common background idea is that people learn in four basic ways:

- visual (learning by seeing);
- verbal or auditory (learning by hearing);
- reading or writing (learning through processing text); and
- kinaesthetic (learning by doing).

In reality, it would be surprising if we do not learn by many of these methods, but some will suit more than others in each particular situation and problem.

One test that has been used quite extensively was developed by Kolb *et al.* (1984) and has subsequently been upgraded several times, with the latest upgrade (Version 4) in 2011 (Kolb and Kolb, 2005, 2013). This test has been extensively researched with considerable technical support available for interpretation. Some examples are available online.

Kolb believed that the learning process involved:

- having a concrete experience;
- this leads on to observation and reflection;
- giving rise to abstract concepts which are mentally held; and
- these are then used in new situations to provide a test of the concept.

If the concept does not work, then the *whole process starts again* until a match between the concrete experience and the concept is arrived at.

Just how important each aspect is to an individual will vary, so that, for example, they might find the concrete experience very important in contrast to, perhaps, being able to work from more abstract representations such as a book description of a problem and solution. Thus, the idea of learning styles was developed. While there is a continuum, four distinct classifications were developed which subsequently, in the latest version of the test (V4), have been refined into nine styles. The original four styles are (source: Kolb Learning Style Inventory, reprinted with permission. For details see the Acknowledgements on p. xv) as follows.

The diverging style

This combines the Concrete Experience and Reflective Observation phases. People with this learning style are best at viewing concrete situations from many different points of view. Their approach to situations is to observe rather than take action. If this is your style, you may enjoy situations that call for generating a wide range of ideas, such as brainstorming sessions. You probably have broad cultural interests and like to gather information. In formal learning situations, you may prefer working in groups to gather information, listening with an open mind and receiving personalized feedback.

The assimilating style

The assimilating style combines the Reflective Observation and Abstract Conceptualization phases. People with this learning style are best at understanding a wide range of information and putting it into concise, logical form. If this is your learning style, you probably are less focused on people and more interested in abstract ideas and concepts. Generally, people with this learning style find it more important that a theory have logical soundness than practical value. In formal learning situations, you may prefer lectures, readings, exploring analytical models and having time to think things through on your own.

The converging style

This style combines the Abstract Conceptualization and Active Experimentation phases.

People with this learning style are best at finding practical uses for ideas and theories. If this is your preferred learning style, you have the ability to solve problems and make decisions based on finding solutions to questions or problems. You would rather deal with technical tasks and problems than with social and interpersonal issues. In formal learning situations, you may prefer experimenting with new ideas, simulations, laboratory assignments and practical applications.

The accommodating style

The accommodating style combines the Active Experimentation and Concrete Experience phases. People with this learning style have the ability to learn primarily from 'hands-on' experience. If this is your style, you probably enjoy carrying out plans and involving yourself in new and challenging experiences. Your tendency may be to act on intuition rather than on logical analysis. In solving problems, you may rely more heavily on people for information than on your own technical analysis. In formal learning situations, you may prefer to work with others to get assignments done, to set goals, to do fieldwork and to test out different approaches to completing a project.

For simplicity, the four types can be abbreviated to concrete experience (CE), reflective observation (RO), abstract conceptualization (AC) and active experimentation (AE).

For V3.1 of the Kolb test, the four learning modes can be combined into two 'scores' that result from combining concrete experience and abstract conceptualization (AC–CE) and combining reflective observation and active experimentation (AE–RO). Parameter (AC–CE) measures the extent to which a person emphasizes abstractness over concreteness, while parameter (AE–RO) measures the relative 'action-over-reflection' emphasis. By combining both parameters, a two-dimensional space is developed, and the four learning styles are further defined as:

- convergent, which emphasizes abstract conceptualization and active experimentation;
- divergent, which emphasizes concrete experience and reflective observation;
- assimilation, which emphasizes abstract conceptualization and reflective observation; and
- accommodative, which emphasizes concrete experience and active experimentation.

Following Kolb's V3.1 system, to work out a manager's learning style requires obtaining his ranking of a series of 12 adjective sets. For each set, the manager must rank them in degrees of correctness for his particular approach, or view of life.

For example, for the adjectives 'inquisitive', 'bored', 'unadventurous' and 'excitable', a manager might decide 'excited' is very much like him and so is scored 4, whereas he is certainly not 'bored' and so this adjective is scored 1, and similarly 'unadventurous' does not normally apply to him and so it is scored 2, leaving three for 'inquisitive' which he certainly is.

By way of an example, a group of managers in New Zealand scored such that 23% were accommodators, 20% assimilators, 23% convergers and, therefore, 34% were divergers. As you would expect, the people requiring concrete experiences for learning dominate. Divergers also use 'reflective observation' in contrast to experimentation, which is perhaps a little surprising for managers. Of course, just how a person learns will not necessarily impact on their managerial ability, but a knowledge of the best way for an individual to learn helps in working on developing a manager's skills. Thus, while concrete experience is dominant, there are still people who can learn in abstract form so that they can improve from reading. Concrete-oriented people, however, will best be helped through demonstrations and practice, though in the end everyone does need to have hands-on experience.

Version 4 of the Kolb test, as noted, has nine styles, thus further subdividing the main styles. These styles are labelled initiating, experiencing, imagining, reflecting, analysing, thinking, deciding, acting and balancing.

It should also be noted that the evidence relating learning modes with outcomes is not particularly strong. However, common sense would suggest that some people do learn more efficiently through certain modes than others. For the case study managers, Margrave is very much an active experimenter/concrete experience person with an AE score of 30 out of a possible maximum of 36, and CE of 30/36. In contrast, his abstract conceptualization (AC) score was 12 and reflective observation (RO) 18, indicating that he 'learns by doing'. For Hank, his scores are much more even across the types with CE 26, AE 24, AC 21 and RO 19. The fact that Hank spends at least an hour a day reading complies with his more abstract learning style compared with Margrave, though he still does score reasonably highly on the 'doing' categories.

The research available on learning is largely based on urban situations so it cannot be categorically concluded that, for example, the Kolb test will definitely be useful in primary production. Some rural managers will find interpreting the adjectives in the Kolb test difficult, so that a test using similar terminology that can be directly related to each domain might well be useful.

In the Kolb concept, experience is an important part of the learning cycle. Even if theorists do not agree with the approach, all would agree that experience and practice are an important part of becoming a good manager. The ability to observe, learn and implement the lessons from experience is probably very important in practical occupations. More is said about experience in the next chapter (Chapter 3).

Margrave, the case study manager, has an interesting learning style. He firmly believes 'reading is the style I learn best from. I read everything I can, particularly during meals when I was young and single. The reading led to thinking of options.' Hank, in contrast, believes he is more of a kinaesthetic learner and needs to get out and practice. He found learning computer skills needed plenty of time at the keyboard in contrast to reading manuals. However, Hank also noted his memory was very good, something everyone aspires to, and for this reason makes sure he reads at least an hour per day to build up his knowledge and obtain ideas as well as critique others. 'There is so much to cover I have little time to do other early-morning and late-night activities. I haven't read a novel in 20 years.'

While the discussion has focused on the Kolb test (Kolb and Kolb, 2005, 2013), as noted there are many learning style tests available. Most involve answering a set of questions which are then scored leading to a conclusion using results from many uses of the tests. Two examples of the kinds of questions are given below. (A search on the World Wide Web (WWW) will provide a reader with many tests. Before use their validity should be checked through the literature; the examples provided were picked at random.)

For these example questions each is scored on a 1 to 4 scale of correctness.

> I feel the best way to remember something is to picture it in my head.
>
> I follow oral directions better than written ones.
>
> I often would rather listen to a lecture than read the material in a textbook.
>
> I am constantly fidgeting (e.g. tapping pen, playing with keys in my pocket).
>
> (For the full test see: www.ldpride.net/learningstyles.MI.htm, accessed 26 August 2024)

And another set of examples is:

> I understand something better after I
>
> (a) try it out.
>
> (b) think it through.
>
> I would rather be considered
>
> (a) realistic.
>
> (b) innovative.
>
> When I think about what I did yesterday, I am most likely to get
>
> (a) a picture.
>
> (b) words.
>
> (For the full test see: https://md.utoronto.ca/sites/default/files/Learning%20Styles%20Inventory.pdf, accessed 10 September 2024)

Clearly the scoring system for each of these examples is different. Thus, the interpretation will depend on the scores obtained relative to a standard set.

Concluding Comments

Thus, the age-old question is – 'Is an excellent manager born or trained?' To answer this question, it is important to understand the components of a good manager, and to consider whether the components can be improved through training of some kind. This chapter has contained discussions on the factors likely to impact on good management, and provided formal written tests that can be used to assess the characteristics of managers. The results enable us to form an opinion on the attributes that relate to good management. The tests cover managerial style (personality) and aptitude (intelligence). Also discussed is a manager's attitude to risky situations as this will influence the form of decisions made, as will a manager's set of objectives. Furthermore, as the effort a manager puts into making good decisions is related to his belief in just how much control he has over business outcomes, a test for assessing his 'locus of control' will be introduced in the next chapter (Chapter 3). Finally, as the objective of studying managerial skill is to improve managers' skill level, a test of a manager's learning style was discussed. Some doubt about its efficacy and value, however, was commented on. Similar comments apply to motivation tests and the concept of 'emotional intelligence'.

To understand skill improvement, it is important to understand the basic characteristics that give rise to skill. The personal traits discussed are the building blocks, so familiarity with them leads to a sound grounding for understanding people and their abilities. This chapter has provided this background and understanding. Another characteristic that has not been mentioned is 'entrepreneurship'. This term is used to describe people who create new ideas for products and production systems, and are able to put the ideas into successful operation. Whether the basic personality traits are highly related to this characteristic is not yet clear, but this is certainly a strong possibility. The openness trait could well give rise to entrepreneurship. Entrepreneurs might also be classed as 'experts', as discussed earlier. However, further research may more directly correlate entrepreneurship with personality, or conclude whether a new trait needs to be added to the management-style traits.

Finally, it is worth reinforcing the comments about the use of normal observation in classifying managers and assessing their current rating in the skill continuum. Careful and continuous observation of how a manager operates should give an astute observer a good representation of a manager's characteristics, especially an observer who fully understands the traits and factors that give rise to high skill. Where formal tests can be an advantage is for people who do not have these judging abilities and, second, as a discussion point when working on improving a manager's skill level. In the end, a combination of tests and observation will be important in most cases.

Chapter 3 looks at quantifying the importance of the factors contributing to managerial skill. With this knowledge it is possible to judge where any improvement effort should be focused.

2A. Managerial Style Test

Tick ONE box that best records your degree of belief in the statements.

1. You tend to mull over decisions before acting.

   TRUE ❑ ❑ ❑ ❑ ❑ NOT TRUE

2. You find it easy to ring up strangers to find out technical information.

   TRUE ❑ ❑ ❑ ❑ ❑ NOT TRUE

3. For most things you seek the views of many people before making changes to your operations.

   TRUE ❑ ❑ ❑ ❑ ❑ NOT TRUE

4. You usually find discussing everything with members of your family and/or colleagues very helpful.

   TRUE ❑ ❑ ❑ ❑ ❑ NOT TRUE

5. Where there are too many jobs for the time available you sometimes become quite anxious.

   TRUE ❑ ❑ ❑ ❑ ❑ NOT TRUE

6. You tend to tolerate mistakes and accidents that occur with employees and/or contractors.

   TRUE ❑ ❑ ❑ ❑ ❑ NOT TRUE

7. You share your successes and failures with neighbouring businesses.

   TRUE ❑ ❑ ❑ ❑ ❑ NOT TRUE

8. Keeping records on just about everything is very important.

   TRUE ❑ ❑ ❑ ❑ ❑ NOT TRUE

9. You admire business/grower colleagues who are financially logical and don't let emotions colour their decisions.

   TRUE ❑ ❑ ❑ ❑ ❑ NOT TRUE

10. You sometimes don't sleep at night worrying about decisions made.

    TRUE ❑ ❑ ❑ ❑ ❑ NOT TRUE

11. You find investigating new business/growing/production methods exhilarating and challenging.

    TRUE ❑ ❑ ❑ ❑ ❑ NOT TRUE

12. You tend to write down options and calculate monetary consequences before deciding.

    TRUE ❑ ❑ ❑ ❑ ❑ NOT TRUE

13. You tend to worry about what others think of your methods.

TRUE ❑ ❑ ❑ ❑ ❑ NOT TRUE

14. You are happy to make do with what materials you have to hand.

TRUE ❑ ❑ ❑ ❑ ❑ NOT TRUE

15. You find talking to others about growing ideas stimulates and excites you as well as increasing your enthusiasm for new ideas.

TRUE ❑ ❑ ❑ ❑ ❑ NOT TRUE

16. Having to make changes to well-established management systems and rules is a real pain.

TRUE ❑ ❑ ❑ ❑ ❑ NOT TRUE

17. You normally don't rest until the job is fully completed.

TRUE ❑ ❑ ❑ ❑ ❑ NOT TRUE

18. You normally enjoy being involved in manager organizations.

TRUE ❑ ❑ ❑ ❑ ❑ NOT TRUE

19. You sometimes believe you are too much of a stickler for checking and double-checking that everything has been carried out satisfactorily.

TRUE ❑ ❑ ❑ ❑ ❑ NOT TRUE

20. When the pressure is on you sometimes become cross and short with others.

TRUE ❑ ❑ ❑ ❑ ❑ NOT TRUE

21. You generally choose conclusions from experience rather than from hunches when they are in conflict.

TRUE ❑ ❑ ❑ ❑ ❑ NOT TRUE

22. You are inclined to let employees/contractors do it their way.

TRUE ❑ ❑ ❑ ❑ ❑ NOT TRUE

23. You not only speak your mind and ask questions at manager/grower meetings, but also enjoy the involvement.

TRUE ❑ ❑ ❑ ❑ ❑ NOT TRUE

24. It is very important to stick to management principles no matter what the pressure to do otherwise.

TRUE ❑ ❑ ❑ ❑ ❑ NOT TRUE

25. You are much happier if everything is planned well ahead of time.

TRUE ❑ ❑ ❑ ❑ ❑ NOT TRUE

2B. Managerial Aptitude Test

This test has been designed for animal production SME businesses with many of the questions containing technical information. Clearly it is important to check for technical proficiency for the specific domain. For any particular business the technical questions will need changing to reflect the technology under consideration. For business 'in general' detailed technical questions will need removing. Also, some questions relate to a particular country and will need consequent modification, and similarly for the current market conditions.

Instructions

- Leave any question blank if you don't know the answer or wish to bypass it.
- Answer by writing in the box the number of the correct answer where choices are given, **OR** the actual answer, **OR** tick the relevant box.

I. Memory

1. How many acres are in a hectare? [acres]

2. For sedimentary soils what is the desirable Olsen P test? [ ]

3. What is a desirable pH for good growth? [ ]

4. What is the normal commission on stock sales? [%]

5. What is *Trifolium repens?*

(1) White clover (2) Lucerne (3) Red clover (4) Wheat [ ]

6. In the RMA, what is a complying activity? [ ]

(1) One where community consultation approves.
(2) One where there are no objectors.
(3) One that is listed in the district plan.
(4) One where the plans must meet the building standards.

7. How many instalments are there for provisional tax? [ ]

8. The Occupational Health and Safety in Employment Act requires a farmer to: [ ]

(1) Keep a register of accidents that harm an employee.
(2) Report all illnesses that keep an employee in bed for more than 1 day.
(3) Put a warning notice on all machines that could cause injury.
(4) None of the above

9. What is the current gift duty rate for gifts less than $27,000/annum? [%]

10. Which of the statements below MOST complies with the efforts to minimize worm resistance to drenches? [ ]

(1) Conduct faecal egg counts.
(2) Rotate drench types.
(3) Rotate the mobs drenched.
(4) Minimize ectoparasites.

11. At the works, Standard superphosphate costs around? [ ]

(1) Mores than $250/tonne.
(2) US$230–249/tonne.
(3) US$210–229/tonne.
(4) Less than US$210/tonne.

II. Experience

1. Think back to a decision you made on feed management (e.g. to buy/sell a significant quantity of hay, to re-grass a paddock, to stop/start irrigation, to use now an area that was shut up for, perhaps hay, or perhaps winter use ...) that, in hindsight, was very wrong.

What was this decision? [ ]

Describe the lessons learnt:

1.
2.
3.

Have you made this, or similar, mistake since or previously? [ ]
Enter Y or N

2. How do you work out the rules to follow when considering when to wean lambs?
Enter the number of the description that is MOST appropriate.
(Read ALL the options BEFORE answering)

(1) The locals and/or neighbours suggested the best rules.
(2) I have discovered from past experience what is best.
(3) I worked out the best rules based on my reading from magazines, books, and field day handouts and the like.
(4) An adviser/consultant told me the rules to follow.
(5) Definitely a combination of most of the above.
(6) Other.

3. Over the years, how much have you changed your management systems as a result of the hard lessons of less-than-hoped-for outcomes.
Tick ONE box to describe the degree of change.

CHANGED A LOT ❑ ❑ ❑ ❑ ❑ NOT CHANGED

III. Creativity

1. Assume your water supply for domestic and stock uses has come from rainwater and a reliable water race. This has been totally adequate. But, the water race system is to be closed down due to some resource consent problems. What do you think are the best two solutions that might be possible and should be investigated?

(1) Get legal advice on the resource consent problem and consider the whole reason for the shutdown – can it be reversed?
(2) Investigate wells and/or stream sources.
(3) Investigate a community water scheme.
(4) Extend the rainwater collection area and storage capacity.
(5) Put in more tanks and truck in water.

No. of best solution

No. of 2nd best solution

2. What business problems would you recommend for research assuming quite limited funds? List, in priority order, the most important topics with respect to a good pay-off to the nation.

1.
2.
3.
4.

3. Assume you have purchased a new block next door to your back boundary that has an identical climate, and good soils. It also has a stream and water right for extensive irrigation. What are you going to do with the new block?

1.
2.
3.
4.

IV. General

1. What is out of place?

(i) (1) Ryegrass (2) Phalaris (3) Alsike (4) Coxsfoot
(5) Chewings fescue

(ii) (1) Aberden Angus (2) Hereford (3) Charolais
(4) Bos taurus (5) Jersey

2. List what you might call your management mistakes, if any, that have occurred over the last 12 months.

1.
2.
3.
4.

3. For my tax records and income tax return I do the following:

Enter the number of the description that is MOST like your practice. (Read ALL the options BEFORE answering).

(1) Prepare the tax return myself using my records.
(2) Write up a cash book of all income/expenses, record the reference number of all source documents, and give the book and the files to my accountant.
(3) Use a computer to record all transactions and give the printout and/or disk to my accountant.
(4) Collect all invoices, statements, sale dockets, etc. and give them to my accountant.
(5) Other.

4. Which statement is INCORRECT?

(1) Escort is for broom. (2) Tordon is for gorse.
(3) Versatil is for scrub. (4) Glyphosate is for grass.

5. A break in wool is caused by ...? (Put the number in the box)

(1) Fungus (2) A night of severe weather (3) Nutritional deficit
(4) Onset of longer days

6. A knapsack is to herbicide as a drenchgun is to ...? (Put the number of the answer in the box)

(1) Anthelmintic (2) Fungicide (3) Sporadicide (4) Innoculum

7. Grandson is to grandfather as ram is to ...? (Put the number of the answer in the box)

(1) Breed upgrade (2) Grand dam (3) Ancestors (4) Progeny

8. What is the next number in the lambing percentage series?

90 95 105 120 ...

9. You are told that the grass cultivar 'smart' is a selection of the cultivar 'slow'. Cultivar 'great' was bred from 'smart'. Thus, we must conclude 'smart' grows faster than 'slow'.

(**T**)rue or (**F**)alse?

10. Jack won some money in a growth rate competition organized by the drench suppliers. Jack spent it ALL in three competing stock and station companies. In the second store he spent US$100 more than half of what he did in the first, AND in the third US$100 more than half of the amount spent in the second. In the first store he bought, of course, US$500 of drench. How much did Jack win?

US$

11. Jack has asked Tom to load the trailer as a prelude to a fencing job. Jack says they need 43 waratahs. Tom can carry five at a time. How many trips did Tom make?

12. If you rearrange the letters TPLSAE you would have the name of a

(1) Sheep breed (2) Clover (3) Grass (4) Fence component

V. Shapes

1. John is working out how to subdivide a very large paddock that has recently been successfully sprayed for a major weed. He is aware separating sunny and dark faces is important, and that stock drift uphill. So far his subdivision looks like:

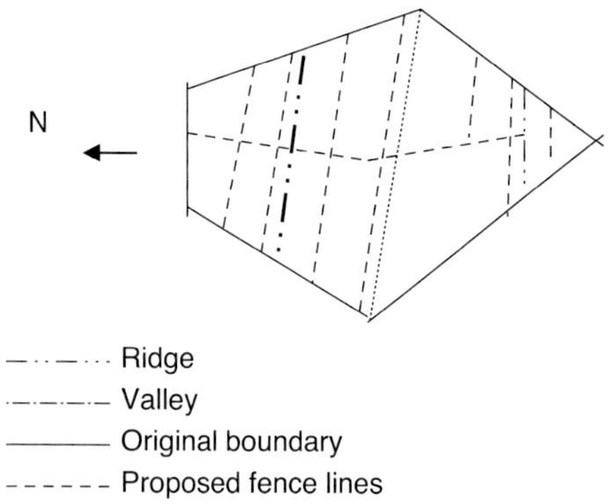

Which shape best fits the shape of the paddocks to go in the area still to be subdivided?

(insert number)

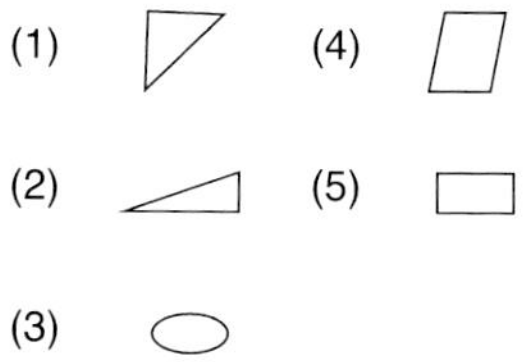

2. Sally has always had a fascination for breeding improved stock through keeping careful records and resultant sire selection. So far Sally has managed to improve the confirmation of the finished lambs quite markedly, as portrayed by the following outline constructed from photos taken of 'on the hoof' dressed lambs at five-year intervals.

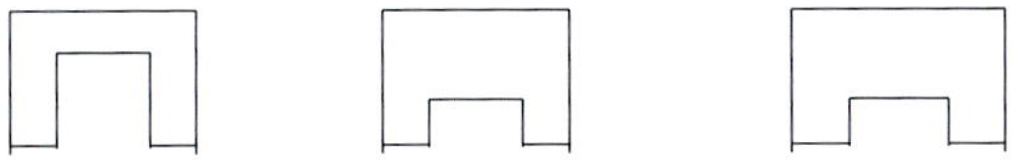

Which of the following outlines best describes what you think Sally will achieve in another five years?

(a) 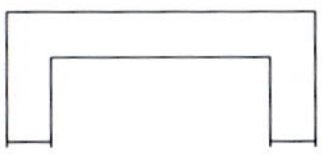 (b) 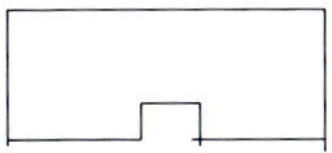 (c) 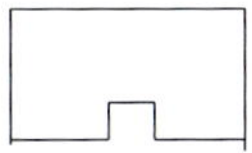

3. Molly is a keen gardener and has put a lot of time into designing, and planting her 'oasis' in the rather isolated place the homestead is located. The plant outline Molly wants between the edge of the front lawn and distant mountains is (Molly selected species accordingly, and won't prune):

Currently the form is:

Which of the forms below would you expect at the halfway stage?

(1) 

(2) 

(3) 

VI. Calculations

1. You receive a call from your diesel delivery person wanting to know how much you need. You know the 1000-litre tank is about 1/4 full. The delivery person comes round about every 2 months, but sometimes it is as much as 3 months at most. While you don't do a lot of tractor work at this time of year you seem to use about 50 litres/week. How much should you order given you always try to never get the tank below 1/5 full?

litres

2. How many ewe lambs are you going to keep? Your flock, just past lambing, is currently 3000 ewes and in two years from now you want 3200 ewes and will NOT buy replacements. In the hogget flock (same number as last year) you have always had 2% deaths and cull about 15% on wool weights. The ewe flock is mixed age and, in the past, you culled 520 ewes per year. Ewe flock deaths average 3%.

Number of ewe lambs to keep?

If the lambing % is 115% S to S, how many ewes should go to the ram for replacements?

3. Drench is on special – the price is the lowest you have seen it for this excellent drench. While you know it has a shelf life of several years and resistance is not expected to be a problem, you reckon it is worth buying a 2-year supply. In the past you have concentrated on a 'clean pasture' policy through rotational grazing and haven't used a lot of drench. In fact, you have only drenched the ewes a couple of times per year, and the hoggets three times per year. The recommended dose is 2 ml/10 kg live weight. Which one of the following ranges covers how much you need to buy for your typical 3000 MA Romney ewes AND replacement flock? (Enter 1, 2, 3, or 4)

(1) <30 litres (2) 30–60 litres (3) 61–90 litres (4) >90 litres

3 The Origins of Managerial Ability

Abstract

An essential part of management is being clear on the business objectives. This chapter gives two basic methods of determining the objectives: observation of the decisions taken, and a formal objective test questionnaire. One such questionnaire is presented and the results of its use for a large sample of managers given. An analysis of the data highlights the main core objectives which cover a range of factors in addition to cash profit. Another factor, a manager's Locus of Control (LOC), is outlined and discussed. LOC reflects a manager's belief in their ability to control outcomes. A questionnaire determining a manager's LOC is provided. The results from a sample of managers give the core factors influencing a manager's LOC. Personality factors are involved among other variables.

With this core of the 'objectives' and LOC, the chapter presents a model of managerial ability also using many other variables thought to be important. Sample manager data is used to determine the importance of each variable on ability as determined by a Structural Equation Model. The chapter ends with highlighting the features of the top managers in the sample.

Introduction

Chapter 2 introduced a range of factors likely to be involved in creating a manager's managerial skill. It is known that a wide range of skill levels exists in any community. This variability can be used to relate these factors to outcomes achieved by using the data from any particular group of managers. The importance of each factor may depend on the environment, but to assess this would require many sets of observations. This chapter, however, contains a discussion on the results of quantifying the relationship between the basic factors and outcomes for a large sample of several types of SMEs in a wide range of environments, thus providing a generalized relationship. The managers in the sample are relatively sophisticated (with, as noted in Chapter 1, approximately a third having some form of formal tertiary education, and certainly all having at least 3 years of secondary education). The SMEs are relatively large in terms of the number of people fully employed relative to worldwide averages, and in terms of the output per person employed. Many would involve a gross investment of at least US$5,000,000.

Determining the importance of each factor enables working out where effort should be directed in improving managerial skill, and, indeed, in determining which factors can perhaps be left aside in such programmes. In assessing the success of a business it is necessary to counter the influence of resource quality to enable comparisons. A primary producing business with high-quality soil, for example, would obviously have higher output per unit investment relative to a poor-quality soil farm, even if the manager was not particularly skilled.

Also relevant is the objective set held by the manager. If a manager has, say, an interest in maximizing leisure, it is not relevant to compare this manager's profit with one with a solely maximum profit objective. Thus, it is important to allow for the different objectives when assessing the origins of managerial ability. A manager maximizing leisure might well be a very efficient manager, but when compared with the profit of other managers s/he may well rank poorly. Thus, ways of measuring the objectives are introduced as a prelude to their inclusion in a quantified model.

Similarly, a part of managerial skill could well be a manager's attitude to how much control s/he believes s/he has over outcomes. Some managers believe, for example, that they have little influence due to the variability and impact of important factors such as the weather and markets. A measure of this control belief is called a manager's 'locus of control' (LOC). A test for measuring this belief is introduced and

 Rural SME Business Management: Human Aspects of Managerial Input (P.L. Nuthall)
DOI: 10.1079/9781800627246.0003

commented upon later. Furthermore, an assessment of how it relates to measures of success is introduced for a sample of managers.

Objectives

As noted, to enable comparisons it is essential to quantify a manager's objectives. Furthermore, it is possible that a manager's objectives will in their own right influence managerial ability. A manager who is quietly content to enjoy production for its intrinsic values might well be content to be an 'average' manager and spend little time improving his skills.

The results of quantifying a set of objectives through a formal questionnaire can be used in conjunction with the casual observations that anyone dealing with a manager will inevitably make. The outcome should be a good assessment of the objectives. Observing the actual decisions made by a manager will indicate the form the manager's objectives take. Furthermore, quantifying the objectives involves writing them down, and this is often a catalyst to reviewing and more carefully defining what the manager is striving to achieve. This process in itself can be beneficial. This is particularly the case where the manager has not formalized a list of objectives, simply preferring to follow their feelings on each issue. This is quite common.

A complicating factor is where there are many owners of a business, or where a family is involved, each one of whom may well have issues with the appropriate objectives to follow. Clearly, a simple case of just one owner involved in the business makes deciding on the objectives very simple. Where this is not the case, it is useful for each interested party to write out their objectives and use the lists as a basis for a round table discussion in a move to obtain a consensus.

There are many formal question sets available in the literature that have been used to assess objectives. Listed below is one example in which the manager is asked to rate the truthfulness associated with a series of statements. It will be noted that all the major aspects that might be of interest to a manager and the family are included.

Question set to assess a manager's goals and aims

Tick ONE box that best records your degree of belief in the statements.

1. It is very important to pass on the business to family members.

   TRUE ❑ ❑ ❑ ❑ ❑ NOT TRUE

2. It is important to earn the respect of managers in the local community.

   TRUE ❑ ❑ ❑ ❑ ❑ NOT TRUE

3. Making a comfortable living is important.

   TRUE ❑ ❑ ❑ ❑ ❑ NOT TRUE

4. It is very necessary to keep debt as low as possible.

   TRUE ❑ ❑ ❑ ❑ ❑ NOT TRUE

5. It is essential to plan for reasonable holidays and plenty of leisure time.

   TRUE ❑ ❑ ❑ ❑ ❑ NOT TRUE

6. Attending field days and manager meetings is vital.

   TRUE ❑ ❑ ❑ ❑ ❑ NOT TRUE

7. It is very important to reduce risk using techniques like diversification, producing conservatively, keeping cash reserves, etc.

   TRUE ❑ ❑ ❑ ❑ ❑ NOT TRUE

8. Developing facilities and systems that give good working conditions is crucial.
TRUE ❑ ❑ ❑ ❑ ❑ NOT TRUE

9. It is very important to ensure employees enjoy their jobs.
TRUE ❑ ❑ ❑ ❑ ❑ NOT TRUE

10. Doing jobs that I enjoy is a very important part of the operation.
TRUE ❑ ❑ ❑ ❑ ❑ NOT TRUE

11. Minimizing pollution is very important.
TRUE ❑ ❑ ❑ ❑ ❑ NOT TRUE

12. I enjoy experimenting with new products and production systems.
TRUE ❑ ❑ ❑ ❑ ❑ NOT TRUE

13. Proper retirement planning is a major consideration.
TRUE ❑ ❑ ❑ ❑ ❑ NOT TRUE

14. You must always be striving to increase the total value of assets.
TRUE ❑ ❑ ❑ ❑ ❑ NOT TRUE

15. Constantly expanding the size of the business is absolutely necessary.
TRUE ❑ ❑ ❑ ❑ ❑ NOT TRUE

16. Aiming for maximum sustainable net cash returns is very important.
TRUE ❑ ❑ ❑ ❑ ❑ NOT TRUE

17. Maintaining a presence in local community activities is important.
TRUE ❑ ❑ ❑ ❑ ❑ NOT TRUE

18. It is very important to improve the condition of the business (fertility, facilities, etc.).
TRUE ❑ ❑ ❑ ❑ ❑ NOT TRUE

19. Giving assets to the children so they can pay for education and/or set up businesses is very important.
TRUE ❑ ❑ ❑ ❑ ❑ NOT TRUE

20. While I don't particularly enjoy managing, I carry on as I don't have a background that allows shifting into another occupation.
TRUE ❑ ❑ ❑ ❑ ❑ NOT TRUE

As an example of the goals held by a particular business community (from a New Zealand sample covering all production types and regions), the results for two sample surveys using this questionnaire are given in Table 3.1. The table contains the mean scores and standard deviations (SD; a measure of the variability around the mean) for the goals and aims listed in the questionnaire. Clearly there is a range within the community. The scale used was 1 to 5, where one records that the goal or aim is 'true' for the respondent through to a five for 'not true'.

The most highly ranked goal is 'making a comfortable living', and this was relatively consistent across most respondents (SD = 0.74 and 0.69 in the two surveys). Statistically, it is expected that 95% of all observations fall within the range of ±1.96 SD. In contrast, the goal of 'constantly expanding the size of the business' was scored the lowest at 3.4 with SD of 1.3, so there is some variation between people.

Table 3.1. Mean score and standard deviation for a range of objectives (1 = 'true' to 5 = 'not true'). Results from two random surveys.

		Survey one		Survey two	
		Mean score	Standard deviation	Mean score	Standard deviation
1.	It is very important to pass on the business to family members	3.15	1.45	2.73	1.46
2.	It is important to earn the respect of managers in the local community	2.66	1.30	2.51	1.25
3.	Making a comfortable living is important	1.47	0.74	1.45	0.69
4.	It is very necessary to keep debt as low as possible	2.26	1.35	2.26	1.38
5.	It is essential to plan for reasonable holidays and plenty of leisure time	2.24	1.20	2.35	1.27
6.	Attending field days and manager meetings is vital	2.84	1.27	2.89	1.30
7.	It is very important to reduce risk using techniques like diversification, producing conservatively, keeping cash reserves, etc.	2.44	1.89	2.38	1.18
8.	Developing facilities and systems that give good working conditions is crucial	1.71	0.81	1.66	0.78
9.	It is very important to ensure employees enjoy their jobs	1.57	0.75	1.47	0.78
10.	Doing jobs that I enjoy is a very important part of the operation	1.83	1.01	1.78	0.98
11.	Minimizing pollution is very important	1.67	0.86	1.69	0.90
12.	I enjoy experimenting with new products and production systems	2.53	1.13	2.51	1.16
13.	Proper retirement planning is a major consideration	2.28	1.20	2.25	1.25
14.	You must always be striving to increase the total value of assets	2.18	1.12	2.20	1.18
15.	Constantly expanding the size of the business is absolutely necessary	3.38	1.31	3.39	1.29
16.	Aiming for maximum sustainable net cash returns is very important	1.92	1.01	2.00	1.07

Continued

Table 3.1. Continued

		Survey one		Survey two	
		Mean score	Standard deviation	Mean score	Standard deviation
17.	Maintaining a presence in local community activities is important	2.79	1.24	2.54	1.23
18.	It is very important to improve the condition of the business (fertility, facilities, etc.)	1.52	0.73	1.48	0.71
19.	Giving assets to the children so they can pay for education and/or set up businesses is very important	2.60	1.22	2.47	1.26
20.	While I don't particularly enjoy managing, I carry on as I don't have a background that allows shifting into another occupation	4.43	1.04	4.51	1.03

While 'reducing risk' is scored between these extremes (2.4), it does have a higher standard deviation (1.9) indicating the respondents' views vary significantly. Other important aims include:

- improving the condition of the business;
- ensuring employees enjoy their jobs;
- minimizing pollution;
- facilities for good working conditions;
- job enjoyment; and
- maximizing sustainable net cash returns.

Note that the latter goal is only seventh on the priority list, though 'making a comfortable living' does appear first. It should also be noted that the mean scores were significantly different (F statistic = 190.25), indicating the priority ranking was meaningful. It is clear managers, on average, tend to be 'way of life' people in contrast to simple profit maximizers. However, each individual manager will have a different ranking for each objective.

In the interests of summarizing the goals and aims, a correlation analysis of the combined surveys was conducted to give the basic factors representing the managers' core objectives and goals (this is called 'factorizing' the variables – for details see any statistical analysis textbook. Essentially it uses correlation calculations to find groups of independent characteristics.)

Factor One featured:

- good facilities;
- enjoyable jobs;
- minimizing pollution; and
- employees' enjoyment (work enjoyment factor).

Factor Two featured profit and assets (the 'wealth' factor).

Factor Three involved the family and community ('people' factor).

Factor Four was about fun and leisure ('leisure' factor).

Factor Five involved debt and risk minimization goals ('risk-aversion' factor).

In summary, the factors are called

- work enjoyment;
- wealth;
- people;
- leisure; and
- risk.

They clearly cover most aspects. Any one manager or family will have an importance ranking for each. This ranking may well influence their attitude to managerial skill improvement.

From the results of case study manager Hank's answers to the questionnaire it was clear he was very interested in the wealth factor as an objective, but he also believed 'family and community' was reasonably important in guiding his decisions. For 'fun and leisure', Hank had an average score as did his 'work enjoyment' factor. His responses reinforced his answers to the risk attitude test with the 'minimize risk' objective featuring quite strongly. It is interesting, and reassuring, to compare his objectives with his actions. Currently Hank is developing an additional business purchased not long ago, thus helping satisfy his wealth objective. He works long hours, but does get involved in community affairs, so again complying with the questionnaire answers.

Margrave responded similarly over the wealth factor. Of course, you would expect excellent and ambitious managers to have this objective. But there again, some excellent managers might well be more interested in family, work enjoyment and community and therefore they must be judged on what their objectives are, not an observer's assumptions on profit maximization. For 'work enjoyment', Margrave had a score in the middle of the road, but he found 'family and community' not particularly important. This probably reflected in part that his spouse was in full-time work in the latter years. 'Fun and leisure' was similarly not very important, reflecting his passion for primary production in contrast to other activities. Finally, his 'minimize risk' score showed this was not a major concern when making decisions. This also reflects his risk attitude.

Note that a further, more detailed examination of objectives, their origins and the relationship between objectives and families is provided in Chapter 9.

Locus of Control

For managers who have little confidence in their ability to control outcomes, the inference is that risky events, such as the weather, the state of the market and the international exchange rate, are more important in the success of their business than the decisions made. For exactly the same environment, however, some managers will believe they have a reasonable degree of control over outcomes. Of course, just where the truth lies will depend on the environment and situation each manager finds her/himself in. Certainly, in situations where the weather, markets and all the other uncertain areas are extremely variable, and good prediction is not possible, and where the number of alternative products and production methods is minimal, the manager will have less control compared with an environment more stable and predictable with a wide range of choices.

Despite the environmental situation, it is useful to know just how much control a manager believes is possible. This belief will impact on efficiency and skill. The reasons a manager holds any one belief will depend on past experiences and the ability to relate decision with outcome. Comparisons with neighbouring businesses also help formulate a conclusion. In addition, the lessons learnt while acquiring managerial skill, and the influence of other managers and parents, will have influenced the attitude.

A test for assessing the degree of control belief is called the 'locus of control' (LOC). A manager is classified as an 'internal' if he believes he has considerable control, in contrast to a manager who has an 'external' LOC. Of course, the situation is not as black and white as this, with a continuum existing from one end of the spectrum to the other. A wide range of question sets to assess the LOC have been developed for a wide range of interest areas (e.g. control of a person's health, workplace safety, motor accident control, and so on). Listed below is a question set designed for SMEs that is based on one of the first general tests developed, but with the terminology and situations changed to relate to managers.

Question set to assess a manager's Locus of Control

For each of the following statements indicate how true it is. Each statement has five boxes beside it – tick only the ONE that best describes your degree of belief in the statement.

1. So far I have managed to largely achieve my goals.

TRUE ❑ ❑ ❑ ❑ ❑ NOT TRUE

2. I never try anything that might not work.

TRUE ❑ ❑ ❑ ❑ ❑ NOT TRUE

3. I'm using exactly the same production methods that I have used for many years as they have stood the test of time.

TRUE ❑ ❑ ❑ ❑ ❑ NOT TRUE

4. It's no use being stubborn about a job or management approach that doesn't initially work.

TRUE ❑ ❑ ❑ ❑ ❑ NOT TRUE

5. I reckon 'good luck' doesn't exist – 'luck' is really good management, and 'bad luck' poor management.

TRUE ❑ ❑ ❑ ❑ ❑ NOT TRUE

6. It is safer not to rely on others to get the job done well and on time.

TRUE ❑ ❑ ❑ ❑ ❑ NOT TRUE

7. I'm able to get others to do the jobs my way.

TRUE ❑ ❑ ❑ ❑ ❑ NOT TRUE

8. Too often I end up having to run my business to suit others' demands.

TRUE ❑ ❑ ❑ ❑ ❑ NOT TRUE

9. While being a good manager involves some training, experience and reading, management skill is mainly determined by your genes.

TRUE ❑ ❑ ❑ ❑ ❑ NOT TRUE

10. You can work hard at creating good relationships between neighbouring managers, but often your efforts fall on deaf ears as people are commonly uncooperative and self-interested.

TRUE ❑ ❑ ❑ ❑ ❑ NOT TRUE

11. I find most employees work hard and finish the tasks set very adequately after a bit of training where necessary.

TRUE ❑ ❑ ❑ ❑ ❑ NOT TRUE

12. The years when the business has shown poor production and profit have been due to circumstances totally out of my control.

TRUE ❑ ❑ ❑ ❑ ❑ NOT TRUE

13. In local body affairs it's easy for a hard-working and dedicated individual to have an impact in getting changes for the better.

TRUE ❑ ❑ ❑ ❑ ❑ NOT TRUE

14. Often I get frustrated as circumstances beyond my control impede the smooth progress of my management plans and decisions.

TRUE ❑ ❑ ❑ ❑ ❑ NOT TRUE

15. Some people seem to be just lucky and everything works out for them, but it hasn't happened to me much.

TRUE ❑ ❑ ❑ ❑ ❑ NOT TRUE

16. I tend to carefully plan ahead to ensure my goals are achieved, and often do budgets and commit my ideas to paper.

TRUE ❑ ❑ ❑ ❑ ❑ NOT TRUE

17. I seldom change my management and production systems unless I'm doubly sure the change will be positive. So much depends on chance.

TRUE ❑ ❑ ❑ ❑ ❑ NOT TRUE

18. When things go wrong it is so often due to events beyond my control – the weather ruins the marketing plans, the produce auction I choose has a sudden price dip, etc.

TRUE ❑ ❑ ❑ ❑ ❑ NOT TRUE

19. When I know I'm right I can be very determined and can make things happen.

TRUE ❑ ❑ ❑ ❑ ❑ NOT TRUE

In developing an overall score for a respondent's belief, the score on each question is added after the score is reversed for the statements where 'true' means low control so a high total score means a strong belief in controlling outcomes. The scores can then be converted to a percentage figure, so a score of 100 indicates a very strong control belief.

As an example of the kind of control scores that are typical over a wide range of rural businesses in a temperate climate (New Zealand), the distribution data obtained from a sample survey is given in Fig. 3.1 together with the normal distribution. The average score was 67%, and the standard deviation 8.5. Clearly a wide range of control beliefs existed in the

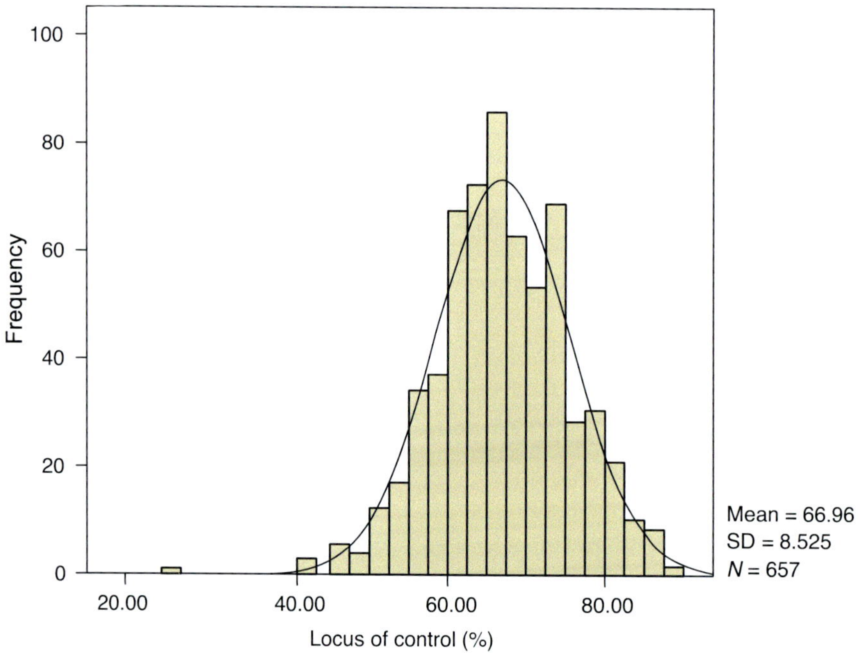

Fig. 3.1. Distribution of the respondents' 'locus of control'.

community, though the majority falls within the 55–80% band.

As there are similarities in some of the control statements, it is useful to look at the correlations between statements to explore whether there are groupings that summarize the basic components of the manager's control belief. This analysis suggested that six factors underlie the beliefs. These can be called:

- beyond control;
- conservative traditionalist;
- determined despite bad luck;
- flexible achiever;
- gene-based traditionalist; and
- careful and determined planner.

Managers with a high proportion of Factor One have little control belief and accept that outcomes are largely luck, whereas people with a high proportion of Factor Two believe you make your own luck through using tried-and-tested methods. Factor Three relates to being able to ride through bad luck, whereas Factor Four involves a belief that it is possible to achieve if you are flexible. In complete contrast, Factor Five embodies an acceptance that your genes determine ability, but despite this good achievement is possible. Finally, Factor Six is the belief that careful planning does work and will have beneficial pay-offs.

Any one manager will have a mix of these factors leading to an overall control belief. If a manager's LOC is low, or excessively high, thought should be given to whether efforts to change it are appropriate. This is achieved by discussing the manager's beliefs with examples of what is possible. Use of case examples to show what is in fact possible may well assist this process of creating a more realistic attitude. An excessively high LOC may lead to excessive risk taking where techniques to reduce risk are not deployed. An example is diversification.

Research has indicated that age does correlate with LOC, with youth tending to overestimate just how much control is possible. As in most things, appropriate experience, and an ability to learn from this experience, is a very valuable part of becoming a skilled manager as will be shown and discussed in following sections.

Margrave, who operates in a very variable environment, has an LOC of 57%, which is somewhat less than the population average of 67%. Yet he is certainly positive in his view of management with constant change to improve the outcomes. Margrave comments that 'with high risk levels you need to be creative' and 'you can mitigate many things if you think out of the square'. Hank, perhaps due to the high level of irrigation, has a score of 73%, expressing the control he certainly does have. This is also reflected in his purchase of extra land with huge investments in infrastructure.

As for the future, Hank asks, 'How can I exert more control?' Hank believes there will be increasing volatility in prices and markets as the years roll by due to the rapid communication systems that now abound. In the past, the happenings in a market a long way from the production centre were hardly noticed, and only affected prices in the next season. However, now manufacturing and supply corporations are watching the daily movements in key markets and altering manager payouts accordingly. The same sort of thing is happening in input markets. A good example is the fuel market, where over 2 weeks the price at the pump might rise two times, and then suddenly decline only to rise again after 10 days. One wonders what the point of all the ups and downs is, but it certainly makes planning rather difficult. Therefore, Hank concludes, the manager must get used to these ups and downs for they will continue for many years. As managers have little control over the world markets, they must be downgrading their views on the degree of control they have.

Relating Locus of Control to Managerial Factors

There have been a number of quantitative studies relating LOC to various aspects of management. They are not totally conclusive on the importance of a manager's LOC.

Overall, it does appear that the LOC is more of an indication of a manager's propensity to be involved in innovations than their actual managerial ability. Furthermore, as you would expect, management style does impact on a manager's LOC, so it could be possible to infer a manager's LOC from an understanding of their personality.

The research results show that managers tending towards 'internality' are more likely to be adventurous in considering innovations and tend to experience improved financial results relative to 'externals'. Similarly, internals are more likely to seek out and use extension activities and, consequently, learn about possible innovations.

It is also suspected that internals will have created more wealth than their external counterparts. This is a consequence of their innovativeness (Kaine *et al.*, 2004).

For the New Zealand data, the managers' LOC was related to their self-assessed managerial ability. The managers were asked to score themselves on a ten-point skill scale relative to their colleagues. It was found that:

$$\text{Skill} = 3.3 + 0.08A - 0.5E + 0.22I + 0.01L$$

where L is the LOC percentage score, A is age in years, E is the level of education on a five-point scale (1 = 'primary education only' through to 5 = 'three or more years of tertiary education') and I is self-assessed intelligence on a four-point scale (1 = 'below average intelligence' through to 4 = 'highly intelligent'). This equation explained 32% of the variance in the skill level, leaving another 68% unexplained. It is interesting to note the negative coefficient on education in contrast to the impact of self-assessed intelligence and age. LOC does have an impact so that, for example, if the LOC percentage changes by 20%, the skill rating changes by 0.2 on the ten-point scale. However, this is not a major impact.

Despite this low impact of LOC on self-assessed skill, a manager's control attitude undoubtedly does impact on choices in that an external is likely to choose alternatives where the manager perceives greater control is likely. Thus, for example, where the economics of irrigation is marginal, an external is likely to proceed with an investment as the irrigation will give more control and surety.

It is also interesting to note the correlations between the LOC and personality measures. In a study of New Zealand managers there was a:

- 26% (32% in another survey) correlation between the LOC and the anxiety personality factor;
- 44% correlation with conscientiousness; and
- 12% (21%) correlation with extraversion.

The other two traits were not significantly correlated other than a 9% (35%) correlation with openness. You would expect these relationships because an 'open' person can be described as 'original, daring and liberal' and is, therefore, likely to believe in having control, as would a person who is an extrovert (sociable, spontaneous), though the degree of control is not as great. On the other hand, an anxious person would have a lack of sense of control, and this is the case with the 26% (32%) correlation in the opposite direction to the others. Given these quite high levels of correlation it could be suggested that a study of managers' personalities would tell you almost as much as a separate LOC test. To further examine these factors, Chapter 9 contains a discussion on variables that might be regarded as giving rise to a manager's LOC.

Modelling Managerial Ability

Factors giving rise to ability

Understanding the factors that give rise to a manager's managerial ability leads to being able to consider how ability might be improved. Furthermore, having data on the quantitative impact of each factor enables concentrating on the most important aspects of ability.

It has been pointed out that, in particular, management style and intelligence are likely to be important determinants of ability. Equally, you would expect experience to be important, particularly where the manager has the attributes to gain benefit from the experiences. Unfortunately, it is difficult to measure experience as there are no simple quantitative measures. Age is certainly a factor as it is more likely for an older manager to have been exposed to potentially beneficial situations. It is easy and important to include age as an explanatory variable.

Also likely to be relevant will be a manager's formal education, both in terms of the number of years, but also the educational success achieved. There is strong evidence from urban business situations that employees' incomes are highly correlated with education. It is also interesting to note that a major determiner of school attainment is family background. With respect to society's investment in schooling, the return for each additional year of education is around 9% (Ashenfelter and Rouse, 1998). This figure, or very similar, is quite common in several studies. You would expect, similarly, the return to increased education, through managerial ability, will be positive.

The other factor likely to impact on ability is the manager's objectives, or, where they are taken into account, the amalgam of the family members' objectives. Thus the earlier discussion on assessing objectives is due to their likely impact on business success, both in the sense of measuring the right outcomes when assessing success, and for their direct impact on efficiency.

In summary, the factors likely to determine a manager's managerial ability are her or his:

- management style (personality);
- intelligence;
- education;
- experience; and
- objectives.

Also possibly relevant are both the family background, and the prevailing factors that might influence skill that are dominant in the society within which the manager grew up and operates. Of course, the manager's family background helps determine her/his educational experience (it was noted above that family background is a major influence on educational success), but also practices such as involving the children in discussions on business decisions. This involvement could impact on the development of managerial skill, as could the involvement in any business physical (practical) work. Similarly, the dominant culture of the time could influence managerial development so, for example, if it was accepted in the community that children were included in any extension meetings, the experiences could have positive impacts. Another example might be whether children were sent off to boarding schools, in contrast to using local schools, and whether they should go to tertiary education, or learn on the job.

Understanding the importance of all these factors will assist in setting up programmes to ensure young managers are as skilled as humanly possible.

Modelling results

It is easy to draw a diagram of the factors that might be important. It should show which factors influence other factors that in turn determine the level of managerial ability. It is another matter, however, to work out just how important each factor is. Figure 3.2 is a diagram representing what factors might be included in a model.

What is required is a knowledge of how important each arrow is in determining each of the factors in the diagram. The direction of the arrows suggests that the background factors impact on the manager's managerial style, intelligence and objectives. These in turn impact on the education and experiences. Finally, in addition to these factors, management style, intelligence, objectives, education and experience all also directly influence the manager's managerial ability.

To put figures on the arrows' 'importance' requires a technique called 'structural equation modelling'. This works out the regression parameters giving the best fit of the data to the model. If you like, a whole series of linear regressions is carried out to find out the equation which best fits the prediction of the variable in each box. Fortunately, computer packages are available to work all this out, so it is simply a matter of defining the form of the model, and then providing the package with the data available for the sample of managers.

In this case, data for most variables (as described in each box) was available for the sample of New Zealand managers described previously and was used to hypothesize, and to calculate the parameters for, a Structural Equation Model of ability. The model is shown in Fig. 3.3.

The ellipses represent variables that are not actually observed, while the boxes

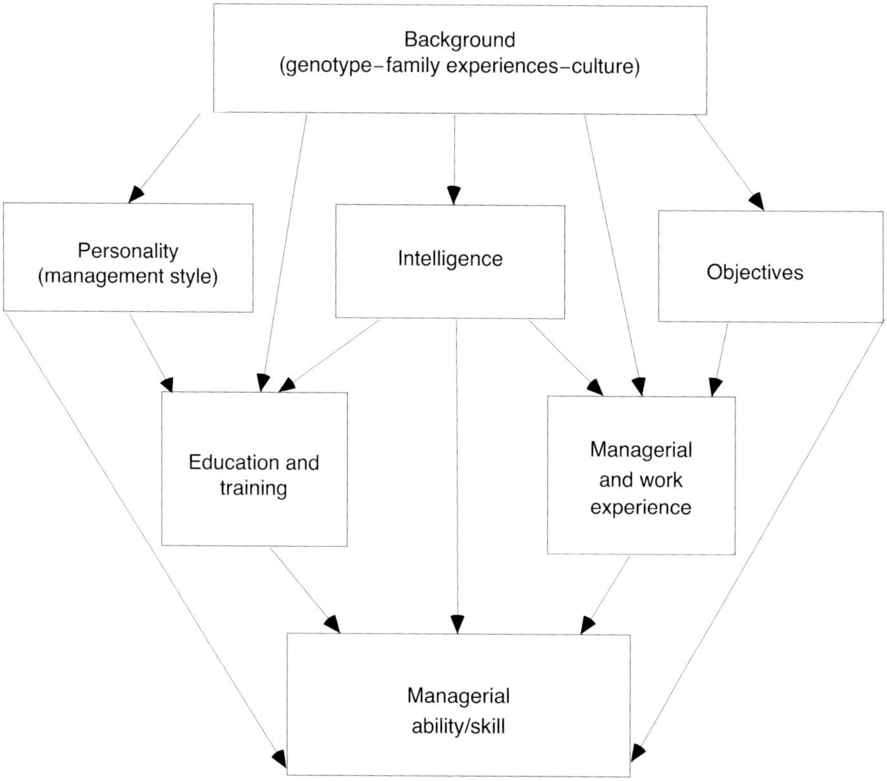

Fig. 3.2. A structural model of managerial ability. (From Nuthall, 2009. See the Acknowledgements, p. xv, for reprint permission details.)

represent the observed variables – in this case, for each manager in the sample. In this case information was available on the managers' parents and forebears as well as their grades for their last year of formal education, highest level of formal education, age, gender, as well as their self-rated managerial ability (or skill). In addition, their locus of control rating was available, as well as the average of their last 5 years' profit increase (sometimes negative) and total asset value increase, and their level of physical output productivity relative to their peers. Given the belief that experience was important, data on a whole range of experience factors was gathered and gave rise to the length of experience (time–experience) variable, and to the value of different types of experience (learn–experience) variable. Finally, information on the managers' management style and objectives, as described previously, was also available.

The model hypothesized that the unobserved 'true ability' is dependent on a manager's management style, the sum of the experience factors (management experience) and the manager's 'true intelligence'. This latter variable was inferred from the manager's grades, level of education and self-rated intelligence. Education, in turn, depends on the manager's parents, age (younger managers had greater opportunities) and experience time. By examining the direction of the arrows, the sources of the basic variables can be inferred. It will be seen that the manager's objectives are assumed to influence a number of variables such as the manager's self-rated ability (skill). Thus, an allowance is made for the nature of the manager in assessing his self-ratings.

The results of the analysis provided standardized regression coefficients for the influence of each impact represented by an arrow. The critical results are the experience → ability arrow with its coefficient of 0.971 relative to

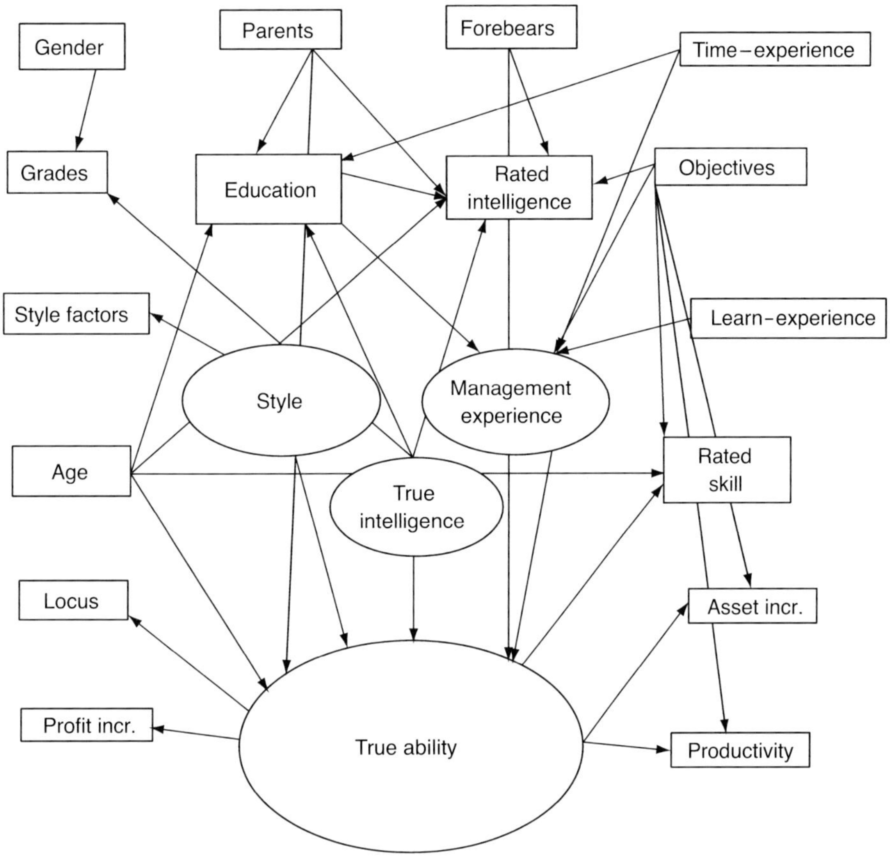

Fig. 3.3. A Structural Equation Model of managerial ability.

the influence of 0.232 for style and 0.109 for true intelligence (the advantage of giving the standardized coefficients is that they provide direct comparability having allowed for the different units of measure used). Thus, experience is approximately four times as important as style in determining managerial ability, and intelligence only a ninth as important. These are important conclusions indicating very clearly the origins of managerial ability. It is worth noting that the second survey data gave similar conclusions. It is important to comment that this does not mean intelligence is not important, for, clearly, to learn the lessons provided by experience a certain degree of education and intelligence is necessary. In addition, it is important to note the measure of success used in the model related to the cash surplus. In work presented later the managers' objectives were widened to include more than just a cash objective. In this case the importance of the direct experience is reduced.

To further reinforce information on the important drivers of managerial ability, the managers in the survey giving rise to the data above were divided into two groups based on their true ability. By comparing the groups, clues as to the most important attributes are provided. The arbitrary division was based on the 70% ability line after converting the managers' scores to a percentage, the distribution of which is given in Fig. 3.4.

Relative to the high-ability managers in the top group, managers who had low ability tended to have a style with:

- high level of anxiety;
- low conscientiousness;

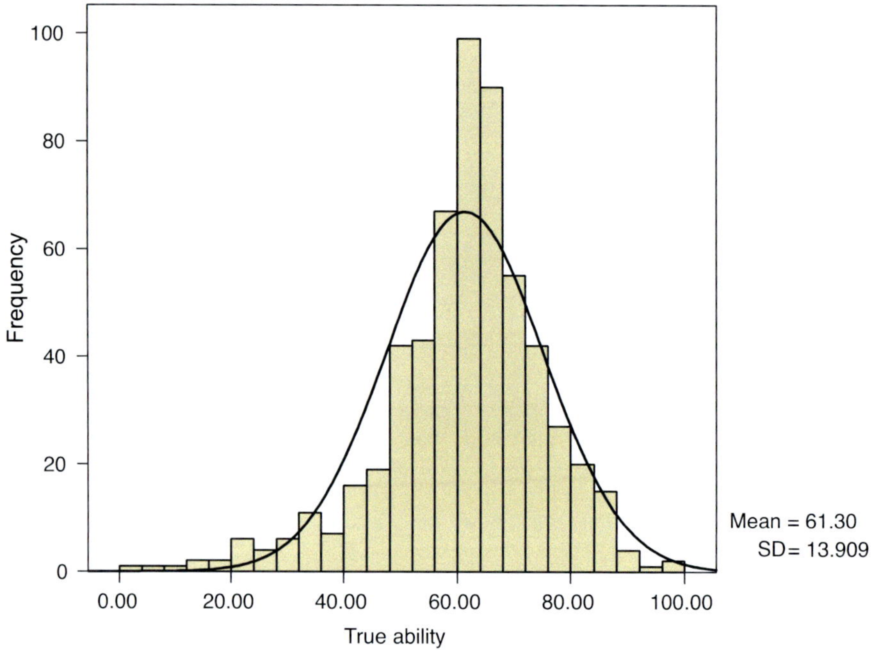

Fig. 3.4. Distribution of managers' true managerial ability. (From Nuthall, 2009. See the Acknowledgements, p. xv, for reprint permission details.)

- low community spirit;
- low levels of consultation with colleagues; and
- a benign outlook.

In addition, those managers who had the following attributes, with respect to their objectives, also tended to be in the low ability group. There were the managers:

- who tended to view their business purely as a profit generator;
- who were not particularly supportive of family; and
- who worked hard at risk reduction.

However, the managers in the high-ability group were those whose parents, for experience, involved them in:

- decision making; and
- the procedures and systems in the business at a young age.

Similarly, the managers were more likely to have high managerial ability if they:

- experienced schooling in rural schools; and
- their parents encouraged developing basic skills like careful observation (such as in Fig. 3.5), good imagination and good people relationships.

For experience at a later stage in life, managers who were slow to learn from their mistakes and experiences tended to have low ability. This leads to the observation that managers who develop good intuition, or tacit knowledge as it is often referred to, tend to be in the high-ability group. Research tends to suggest intuition is not something that some managers have, others not, but an attribute that develops from careful observation of experiences and outcomes. This is explained fully in Chapter 7.

Overall, the model explained 76% of the variability in self-rated managerial skill, so it cannot be taken as a full explanation of ability. Other factors must also be important. Furthermore, some of the measurements used were probably not a perfect reflection of the variables. Despite these cautions, the data does give a strong message about the important factors in ability.

Fig. 3.5. The evidence suggests exposure to relevant management issues at a young age confers high ability.

Clearly family influences on management ability are very important, both for their direct effect, as well as their influence on management style and intelligence. Management style is also important per se, as are the manager's objectives.

On the other hand, intelligence and education are not as important as you might expect, but experience is probably rather more important than anticipated. The implications of these conclusions are discussed further in a later chapter (Chapter 10).

Of course, the figures in this model come from one sample of managers in a developed country situation. While the importance of each factor is probably similar in other countries, it could well be the case that the specific coefficients will be different. Each situation must be assessed on its own merits. Details of the study forming the basis of this discussion on the origins of ability are given in Nuthall (2009).

It is interesting to note that the case study managers largely comprise the factors associated with high managerial ability. They both have considerable experience, though Margrave probably follows a slightly stronger line of SME ancestors relative to Hank. Of course, if you go back far enough, most people have SME ancestors as rural business was the dominant activity. Hank had an extensive early-life management exposure, but while Margrave lived on an SME as a youngster, he was not involved in the management discussions to any great extent.

In addition, when it comes to schooling, Hank scores high on rural experience, whereas Margrave is halfway with a rural primary education, but also with urban secondary education.

Hank was also strongly encouraged to develop his observational skills and imagination as a child, as well as people skills. Margrave, perhaps, did not have quite the same advantages.

Overall, both ended up as being highly skilled managers, no doubt strongly supported by their personalities and sets of objectives that certainly had the right mix, being similar to the managers with high managerial ability in the sample population.

Concluding Comments

It will be noted that the LOC variable was not particularly important in the model. This was due in part to the correlations between LOC and components of management style. It was noted earlier, of course, that LOC probably has more significance in studying a manager and his operations.

Of course, understanding the factors that make a good manager is only half the story. The next step is to consider how the factors might be improved. Clearly, for any one manager they currently have a defined set of parameter values resulting from historic events, so the challenge is to consider if the impacts of history can be altered. This is most unlikely to occur without considerable effort on the part of the manager together with appropriate support. As noted, all these considerations are left to a later chapter (Chapter 10).

The variables likely to bring about change are management style, and the lessons learnt from experience. Intelligence and education are attributes and events from the past that are fixed for all times.

To consider the lessons from experience, it is necessary to think about the important skills learnt. A later chapter (Chapter 5) considers the 'competencies' that are important in good management. Also relevant are the processes used in making decisions. Some lead to good decisions, and some can be biased. Thus, Chapter 4 contains a discussion on the decision processes that are possible, and Chapter 5 contains thoughts about typical biases managers might exhibit.

In that good management is related to the resources a manager holds, Appendix 3A at the end of this chapter presents a set of questions

Fig. 3.6. Even in the animal world parental influence has lasting impacts.

that might be used by a consultant to ensure the manager is totally and fully aware of what they are working with. So often a manager, despite his overall skills, is not totally familiar with her or his real situation, and the difference between reality and wishful thinking can be devastating. Clearly, there is an interaction with the business situation and the desirable managerial attributes. This needs to be recorded.

Anyone interested in good management must constantly be searching for keys and what can be done to turn them. In this sense, while the past cannot be directly revisited, its impact can be mitigated, and possibly reinforced. Parental influences are clearly important (Fig. 3.6). Thus, talking to managers about their origins is always important. Margrave tells a fascinating story of his upbringing with rural primary education and an urban boarding school for the secondary years. He talks about his fascination for primary production, but this only took effect once he left school at 16. He returned to the family home and SME business, but it was soon clear that it was not big enough for two leading protagonists, his father and himself, as they did not always agree on what was needed. This led to Margrave working in other businesses for several years, and then he returned home again only to find the relationship still did not work. The upshot was that his father and mother left the business for retirement to a local town, leaving Margrave to fend for himself. He borrowed 100% of the money necessary that in turn funded his parents' house and living in town. Just after this, there was a major downturn in prices and conditions that eventually led to the business being sold. Somehow, through all this Margrave kept learning and ended up as an excellent manager. Some might say the experience, and his passion to read and learn, together with an appropriate management style, all led to a manager that produced successful outcomes. Experience, provided the lessons on offer are learnt, is a wonderful trainer – as clearly shown by the modelling results referred to above.

However, Margrave noted, 'Rarely do the same things happen twice so that experience is not so useful in a direct way, but it provides general lessons that are crucial to future decisions.' One big mistake he learnt from followed long-term weather forecasts for an extremely dry summer. From the previous few years Margrave commented, 'One mistake I always made was not selling assets requiring constant inputs in a dry period soon enough. I was always an optimist.' Thus, currently, with its bad forecast, he decided to sell early. But, 'As the trucks taking the assets rolled out of the business, the rains started and didn't let up all summer.' Experience for you.

3A. On Observation: Preliminary Stocktake – Management/Business Assets Report

Introduction

To plan and make decisions, it is essential that managers know what resources they are working with. Managers are probably fairly confident that they are familiar with the *current state* of their business and finances. Their resources, and their conditions, form the basis from which managers must operate today, tomorrow and every moment into the future. But just how well do they really know their business's state *today*? How good are their observational skills of the recent past, and their memory?

For the example of a primary producing SME, listed below is a set of questions designed to ensure managers, and their advisors, are totally familiar with the resources they have to work with. Managerial skill involves good observational skills, so no matter how good a manager's general skills, if the manager is not conversant with the 'current state' of the business, any decisions made are likely to be suboptimal. For other types of SMEs the topics will need changing to suit, ensuring that all important aspects are covered.

Test – reality check

The example given below is for a business producing wool and meat. For any particular business the questions will need to be altered to reflect the particular business.

For this particular example technology, get the manager to answer the following questions about her/his property in terms of its current situation. Not all the questions will be applicable to each business situation, so some questions should be skipped. In some cases, additional questions might be added to better suit the type of business. The manager should jot the answers on a piece of paper for later checking, or on a photocopy of the question set.

1. EWES
   Divide your flock, including rising 2-year-olds, into THREE live weight groupings and give:

   (i) Lightest group – less than _____ kg/head　　number _____
   (ii) Middle group – between _____ and _____ kg　　number _____
   (iii) Heaviest group – greater than _____ kg　　number _____
   TOTAL _____

2. HOGGETS (young stock) (if none on hand at the moment, leave blank)
Similarly, divide them into TWO groups and give:
(i) Lightest group – less than _____ kg/head number _____
(ii) Heaviest group – greater than _____ kg number _____
TOTAL _____

3. LAMBS (if none on hand, leave blank)
Similarly, divide them into THREE groups in give:

(i) Lightest group – less than _____ kg/head number _____
(ii) Middle group – between _____ and _____ kg number _____
(iii) Heaviest group – greater than _____ kg number _____
TOTAL _____

(iv) average age (weeks) _____

4. OTHER SHEEP (if none on hand at the moment, leave blank)
Similarly, divide them into TWO groups and give:

(i) Lightest group – less than _____ kg/head number _____
(ii) Heaviest group – greater than _____ kg number _____

TOTAL _____

5. COWS
Divide your herd into THREE live weight groupings and give:

(i) Lightest group – less than _____ kg/head number _____
(ii) Middle group – between _____ and _____ kg number _____
(iii) Heaviest group – greater than _____ kg number _____
TOTAL _____

6. HEIFERS (young stock) – less than yearlings
Divide your herd into THREE live weight groupings and give:

(i) Lightest group – less than _____ kg/head number _____
(ii) Middle group – between _____ and _____ kg number _____
(iii) Heaviest group – greater than _____ kg number _____
TOTAL _____

7. HEIFERS – greater than yearlings
Divide your herd into THREE live weight groupings and give:

(i) Lightest group – less than _____ kg/head number _____
(ii) Middle group – between _____ and _____ kg number _____
(iii) Heaviest group – greater than _____ kg number _____
TOTAL _____

8. CALVES (if none on hand, leave blank)
Similarly, divide them into THREE groups and give:

(i) Lightest group – less than _____ kg/head number _____
(ii) Middle group – between _____ and _____ kg number _____
(iii) Heaviest group – greater than _____ kg number _____
TOTAL _____

(iv) Average age (weeks) _____

9. GROWING CATTLE – less than 18 months (if none on hand, leave blank)
Similarly, divide them into THREE groups and give:
(i) Lightest group – less than _____ kg/head number _____
(ii) Middle group – between _____ and _____ kg number _____
(iii) Heaviest group – greater than _____ kg number _____
TOTAL _____
(iv) Average age (weeks) _____

10. GROWING CATTLE – greater than 18 months (if none on hand, leave blank)
Similarly, divide them into THREE groups and give:
(i) Lightest group – less than_____ kg/head number _____
(ii) Middle group – between _____ and _____ kg number _____
(iii) Heaviest group – greater than_____ kg number _____
TOTAL _____
(iv) Average age (weeks) _____

11. OTHER STOCK
(i) Type _______ number ________
(ii) Type _______ number ________
(iii) Type ______ number ________

12. TOTAL SHEEP IS? _______ (add up values above). Is the total correct?

13. TOTAL CATTLE IS? _______ (add up values above). Is the total correct?

14. PASTURE
Divide your grazing pasture into four groups* based on their dry matter cover (make your best guesstimate):
(i) Worst _____ ha of _____ kg DM/ha
(ii) Basic _____ ha of _____ kg DM/ha
(iii) Reasonable _____ ha of _____ kg DM/ha
(iv) Best _____ ha of _____ kg DM/ha
* Include any areas shut up.

15. LUCERNE (ALFALFA) (leave blank if none)
Divide your lucerne (alfalfa) into four groups* based on their dry matter cover (make your best guesstimate):
(i) Worst _____ ha of _____ kg DM/ha
(ii) Basic _____ ha of _____ kg DM/ha
(iii) Reasonable _____ ha of _____ kg DM/ha
(iv) Best _____ ha of _____ kg DM/ha
* Include any areas shut up.

16. FEED CROPS (leave blank if none)
Divide your feed crops into four groups* based on their dry matter cover (make your best guesstimate):
(i) Worst _____ ha of _____ kg DM/ha
(ii) Reasonable _____ ha of _____ kg DM/ha
(iii) Best _____ ha of _____ kg DM/ha
* Include any areas shut up.

17. CASH CROPS (leave blank if none)
Divide your cash crops into two groups:

(i) Looking like average or better yield _____ ha
(ii) Looking like less than average yield _____ ha

18. TOTAL PRODUCTIVE AREA IS? _____ ha (add up values above). Is this total correct?

19. What is your most productive grass variety/cultivar? ____________________

20. What is your most productive clover variety/cultivar? ____________________

21. HAY RESERVES
Divide any hay/baleage into three quality levels and give:

	Dominant species
No. of bales (average _____ kg/bale) of high-quality feed	______________
No. of bales (average _____ kg/bale) of medium-quality feed	______________
No. of bales (average _____ kg/bale) of low-quality feed	______________
How many bales are not adequately covered?	______________

22. SILAGE RESERVES (leave blank if none)
Divide any silage into three quality levels and give:

	Dominant species
No. of tonnes of high-quality feed __________	______________
No. of tonnes of medium-quality feed __________	______________
No. of tonnes of low-quality feed __________	______________

23. STRAW RESERVES (leave blank if none)

How many bales (average _____ kg/bale) of straw do you hold as a feed source?

____

24. CONCENTRATE/GRAIN (leave blank if none)

How many tonnes of concentrate/grain are held as a feed reserve?

Tonnes ____________ Dominant type ___________

25. FENCES (leave blank if not applicable)

Average size of paddocks/fields _____ ha

Biggest	_____ ha
Smallest	_____ ha
Most common	_____ ha

Distance of good-quality fences ______ km

Distance of average-quality fences ______ km

Distance of poor-quality fences ______ km

Main type of fence (describe, e.g. posts & netting) ____________________

No. of paddocks/fields with good shelter ______

average shelter ______

poor shelter ______

No. of paddocks/fields with good water ______

average water ______

poor water ______

Overall reliability of water supply (good, OK, poor) __________

Current average 'fullness' of dams/tanks ______ %

26. BUILDINGS
Wool shed:

No. of stands ______ No. of sheep under cover ______

Condition (poor, OK, excellent) __________

Area (m^2) ______

Sheds:

Type ______ Area (m^2) _____ Condition (poor, OK, excellent) ______

Type ______ Area (m^2) _____ Condition (poor, OK, excellent) ______

Type ______ Area (m^2) _____ Condition (poor, OK, excellent) ______

Hay barns:

No. _____ Total capacity _____ bales of _____ kg/bale

Average condition (poor, OK, excellent) ______

Yards/enclosures:

Sheep – Capacity _____ sheep Condition (poor, OK, excellent) ______

Cattle – Capacity _____ head Condition (poor, OK, excellent) ______

27. MACHINERY (leave blank if not applicable)

Tractor one – Wattage _____ Age _____ years

Condition (good, average, poor) ______

Tractor two – Wattage _____ Age _____ years

Condition (good, average, poor) ______

Tractor three – Wattage_____ Age _____ years

Condition (good, average, poor) ______

Truck/s – Average capacity ______ tonnes

Header/harvester – Size ______ metres cut

Baler – Size _____ (kg/bale) Age _____ years

Condition (good, average, poor) ______

No. of four-wheeler bikes _____ Condition (good, average, poor) ______

No. of items of cultivation machinery _____

Value of the irrigation plant (if any) US$ _____

28. PHYSICAL

Area flat land _____ ha of ___________ soil type

Current available soil moisture level _____ %

Area rolling but cultivable ______ha of ____________ soil type

Current available soil moisture level _____ %

Area hill but NOT cultivable _____ ha of ____________ soil type

Current available soil moisture level _____ %

What area is irrigable? _____ ha

What area do you irrigate? _______________

How much water per hectare? ________

Rainfall:

Average (mm)	spring _____ summer _____	autumn (fall) _____ winter _____
Highest might expect (mm)	spring _____ summer _____	autumn (fall) _____ winter _____
Lowest might expect (mm)	spring _____ summer _____	autumn (fall) _____ winter _____

How much rain (mm) so far this year?

29. PASTURE AND LUCERNE (ALFALFA) PRODUCTION

On average, what do you expect from your pasture/lucerne (alfalfa)?

	Pasture	Lucerne (alfalfa)
Good season ______ kg DM/year	_______	_______
Typical season ______ kg DM/year	_______	_______
Poor season ______ kg DM/year	_______	_______
Highest growth time? __________ (month)	_____ kg DM/day	
Lowest growth time? ___________(month)	_____ kg DM/day	
What is your current stocking rate?	_____ SU/ha	
What do you think is the maximum possible?	_____ SU/ha	

An 'SU' (stock unit) is the feed necessary to support a 50 kg ewe.

30. 'ON-HAND' MATERIAL

How much fertilizer of all kinds is currently held? _____ tonnes

How much grass/clover seed is on hand? _____ kilograms

How much spray material (all kinds) on hand? _____ litres

How much drench of various kinds is on hand? _____ litres

How many kg of wool is 'on hand' waiting sale? _____ kilograms

How much grain/seed in silos is waiting sale? _____ tonnes

31. FINANCIAL

What is the net balance of all your trading bank balances? _____ credit/deficit

What money is currently owed from invoices/statements? _____

What money is currently owed to you? _____ (credits held)

Mortgages and Loans – what is the current outstanding debt for the farm as a whole? __________

For each loan give amount, amount outstanding, principal repayment/year, interest/year* and total payment/year (in US$, perhaps)

Loan one ______________________________

Loan two ______________________________

Loan three ______________________________

Loan four ______________________________

Loan five ______________________________

* Note – if you don't know the split between principal and interest just put in the total payment/year.

32. ASSET VALUE – for the whole business unit (change currency as required)

What is the total current market value of your land and buildings? US$ _____

What is the total current market value of your machinery & plant? US$ _____

What is the total current market value of your sheep? US$ _____

What is the total current market value of your cattle? US$ _____

What is the total current market value of all other assets? US$ _____

33. NET WORTH (change currency as required)

Your current net worth is (sum of assets minus sum of loans and debts outstanding) US$ _____

If any, how much more do you think you could borrow if necessary? US$ _____

What proportion of the amount, if any, would you be happy to borrow? ____ %

How many of the applicable questions could the manager answer with certainty?
And how many was the manager only partially sure about?
And for just how many did the manager take what might be called a guess?

If the manager does not know with reasonable accuracy just what situation the business and assets or debts are in, it is not possible to make the best plans. Good management starts from observation and a good knowledge of the current state of the business.

4 Decision Processes and Goals

Abstract

Managers have choices over the methods they use in making decisions. This chapter outlines a range of approaches typical of common systems. When a manager finds an issue the decision process starts. This might take a linear form of several steps, and use logical decision rules from, for example, production economics. Benchmarks are also commonly used, providing keys to deciding on whether a problem exists. Alternatively, a dynamic approach may prevail. This is a constantly varying system involving continuous reviews and replanning. As theoretically correct systems are time-consuming, sensible shortcuts are often taken. This might involve accepting solutions with 'adequate' output in contrast to optimal output or using intuition as a very common system relying on brain-held decision rules or 'constructs'. These evolve through experience and learning. Good managers have successful rules using certain attributes, making them 'experts'. Emerging technologies must constantly be assessed (theory of planned behaviour) and all decisions should allow for objectives and risk.

Introduction

While each manager will have her/his own particular way of making decisions, there are some common themes. Understanding the possibilities assists in following why a manager makes a specific decision, and in discussing how the decision process might be improved.

Generally, a manager observes factors from the surrounding environment, processes this information and comes to a conclusion about what actions to take, one of which might be to do nothing just yet. Thus the following sequence occurs on a daily basis, irrespective of whether the decision horizon is a few hours or several years:

Stimuli → sensory systems → short-term memory → long-term memory conscious processing ← retrieval subconscious processing ← retrieval

Many of the stimuli being received are acted on automatically, so, for example, we avoid stepping in a hole in the road without realizing we have sorted this little problem. Other stimuli require conscious thought (e.g. a bad storm is forecast) with the information being processed to provide conclusion to, say, take positive action (e.g. shift the machinery under cover; immediately contact the airline to see if product delivery might be impacted).

Some stimuli get considered, and then sent to long-term memory if sufficiently important. The information can then be retrieved when required. For example, perhaps a newspaper reports that the normal spring rainfall in a country producing a significant quantity of wheat for baking flour has not occurred. This information might be stored as something to act on later if further evidence comes to hand that the wheat production is, worldwide, lower than normal.

When talking about managerial ability, our concern is with the success of the decision process used by a manager, particularly the observation systems, and the subsequent conscious and subconscious processing systems. If all these operations work appropriately, the result is good management. As implied, also relevant to the process is the manager's memory storage and retrieval system.

Appreciating the existence of all these systems helps understand good management, and what procedures might be followed to improve a manager's managerial ability.

Besides the decision processes used, the goals and objectives are also important for they

DOI: 10.1079/9781800627246.0004

refine the details of the appropriate processes. If the objective is to simply maximize profit, then the processes in deciding suitable actions should be different from those in situations where not only cash income is important, but also, say, the risk level experienced in earning the income. This chapter contains a discussion on not only the processes that might be used in the various decision stages, but also on their relationship to a manager's goals.

The start-up of a decision process is stimulated by the manager recognizing that some sort of problem exists. This word 'problem' is used in its general sense to define a situation where some kind of thought and analysis is required. Thus, for example, when a new genetically modified yeast is available, there is a problem in the sense that a decision is required on whether it should be used in preference to the current yeast.

Also, of course, problems in the normal sense of the word require sorting – so, for example, when a harvesting contractor is not keeping up with demand, a decision on whether additional labour, and perhaps additional machinery, should be applied is necessary. Whatever the case, successful observation, analysis and decision making allows redirecting the business on to the optimal path.

The simplest decision process, which is not uncommon, is the 'do as the others do' approach. Quite often in any business community some participants are regarded as being the leading managers, so the others watch what they are doing (by direct observation, by talking to the leading manager(s) or perhaps talking to other members of the community who know what is happening). Management then consists of simply 'following the leader'. Whether this succeeds obviously depends on how good the leaders are, whether their situations are comparable, whether the implementation abilities of the followers are suitable for the system adopted and whether their objectives are the same. Having all these attributes occurring at the same time is not common.

The other extreme is where a manager diligently researches every decision situation and has the skills and knowledge necessary to successfully carry out a thorough and logically correct analysis, and then the ability to follow it through. In between these extremes is a range of processes which will be outlined. Of course, in reality there is an infinite range of approaches taken by managers, with a continuum occurring between the extremes. Only a sample set can be summarized. Chapter 7 contains a figure expressing the continuum and simplifications.

What follows is a synopsis of the main processes and how they interact with the objectives. It is assumed that a problem has been identified, so it is then a matter of sorting out its structure, gathering data, analysis, decision, action, feedback and starting the whole process again.

Decision Processes

Linear approach

One theory is the standard rational approach of carrying out a number of steps in a simple linear order. A common list of tasks is:

- information gathering;
- data evaluation;
- problem structuring;
- hypothesis generation;
- preference specifications;
- further analysis and action selection; and
- decision evaluation.

There are many variations on the tasks and ordering that appear in the literature; but suffice to note that the concept is that of a formal set of operations carried out in sequence until a decision is made, which is then implemented. The idea is that a hypothesis is only generated once the data has been collected and examined, so it is assumed that this hypothesis contains the list of likely alternative conclusions. These are then ranked on the basis of their contribution to the objectives. Others might suggest that a hypothesis, or idea, is proposed right at the beginning for it is only once some structure is sorted that you know what data to collect. In fact, it is likely very few people act entirely rationally as is suggested with this linear process.

It was noted earlier that short-term memory has limited capacity to hold material. Thus, in working through the linear process it is necessary for subsection conclusions to be sent off to long-term memory so that the process proceeds in sections with all the information being finally brought together.

The level of success improves with experience in that practice, among other benefits, develops a knowledge of how to break analyses into sections. There is evidence to suggest that in rural SMEs, with all the disciplines involved, many problems are indeed complex and they get broken into simplified chunks. The ability to sort out just what are the key issues, or chunks, is clearly important for success.

For example, decisions about the manufacturing and supply of phosphate fertilizer, in theory, require a knowledge of:

- soil chemistry;
- nutrient distribution patterns in the soil;
- plant requirements;
- soil organic matter;
- water patterns and soil water-holding capacity; and eventually
- response rates and related plant nutrient requirements.

And no doubt there are more aspects than these that are also important. However, it is possible to simplify the whole analysis if it is realized that soil colloids bind phosphorus which not only restricts its leaching, but also provides a buffer of the element that is released over many years.

Such simplifications that still allow a correct analysis and decision to occur provide help in actually reaching a rational decision, for there is a clear limit to the brain power of all decision makers. Indeed, a feature of 'experts' is that they are able to sort out the critical parameters and thus discard all other information that might confuse the issue.

Decision rules

No matter what the approach, linear or otherwise, the manager's knowledge of the appropriate decision rules is crucial. A study of production economics allows working out what rules to use in making an optimal decision.

Research has shown that most production follows a sigmoid curve: as more inputs are added, production increases at an increasing rate, but with even more input, this increase declines to provide decreasing returns. Eventually the shape of the curve describing output to input starts to bend backwards, indicating output is actually falling as more and more input is added. For example, if a fixed set of machinery is used constantly to process raw material without down time to allow proper maintenance, eventually breakdowns occur and, at the extreme, production levels per day will drop.

Under this classical production function (as the relationship between input and output is called), where the output price is sufficient to cover the basic costs so some level of production is economic, it can be proved that the decision rule for maximizing profit is to:

- move along the production function until the marginal cost of the last unit of input added is just equal to the marginal return from the increased output from this last unit of input (Decision rule 1).

Due to the diminishing returns, going past this point means the increased cost of the extra output is not covered by the resultant increase in production. For a full explanation, any standard text on production economics can be referred to, or a basic business management text (Nuthall, 2017).

This rule tells 'how much', per production unit (e.g. machinery process set-up), to produce. There is also the question of:

- what mix of inputs to use (e.g. raw materials); and
- what mix of products (e.g. fertilizer types) to produce.

As you would expect from the description above, production economics 'derives' the decision rules that guide these decisions for optimizing an objective, provided it can be measured. Assuming an objective of maximum cash profit, inputs should be substituted for each other to the point where:

- the marginal rate of substitution between them multiplied by their cost is just equal (Decision rule 2).

The marginal rate of substitution is the amount of, say, a fertilizer raw material that is required to keep output at the same level as it is substituted by another input, say a different source of raw material. If, for example,

1 kg of one fertilizer raw material could be substituted by 1.5 kg of a different raw material with production of the finished product having the same qualities, and the cost of raw material one was US$0.06 /kg while the second raw material costs US$0.02 /kg, it would pay to use more of the second input at the expense of the first as the marginal cost of the first raw material is US$0.06 relative to 1.5 × US$0.02 (US$0.03) for the second. Due to diminishing returns, as you use more of the second material of which the tonnage will be greater, eventually the quantities will need changing (greater demand on the machinery and associated costs) such that at the marginal substitution times the cost is equal, indicating that the least-cost way of production has been determined.

Similarly, the decision rule in deciding the best mix of fertilizer product types to produce is to find the combination such that:

- the marginal value of each product (MVP) is identical (Decision rule 3).

This MVP is the change in physical output as one product is increased at the expense of another, multiplied by its price. Thus, for example, as an extra kilogram of type one (say a nitrogen fertilizer) is produced, this might mean type two (say phosphatic fertilizer) must be decreased by, say, 1.4 kg to release the inputs necessary to produce the extra kilogram of type one. Now, say type one is worth US$0.03/kg and type two US$0.25/kg, then the MVP of type one is US$0.03 and of type two US$0.35. Thus, their MVPs are not equal, so it would pay to produce more type two at the expense of type one. As this shift occurs, the MVP of type two will decline, and of type one will increase. The optimal point that maximizes profit will be where the shift is made until the MVP of each is the same. Similarly for all the other products that might be produced: when all of their MVPs are equal, no change will increase profit.

With these three decision rules it is possible to determine:

- what mix of products to produce;
- what mix of inputs to produce each product (least-cost combination of inputs); and
- how much to produce of each product per technical unit (say labour unit).

Thus, any manager who is making the basic production decisions must understand these rules and do the sums accordingly. These rules apply whether the manager is using a linear approach to the decisions, or some other decision process. While the linear approach tends to focus on formal and significant decisions, such as whether to produce a new product, it is also useable for simple day-to-day decisions such as whether to employ additional labour today. It is still necessary to have the relevant input–output information such as the expected output per day for each level of qualified labour, the cost for each level and information on their reliability and availability, and to apply the relevant decision criteria such as whether the marginal cost of hiring each type relative to the marginal return of the increased output does in fact provide a surplus, and similarly for all the different skill levels available.

While optimal decision rules are clearly established, many managers use inappropriate rules such as, for example, using average returns and costs instead of their marginal counterparts. Their conclusion will be biased.

Problem recognition

Of course, it is worth restating that the success of the decision process depends on the manager identifying the existence of a decision problem in the first place. In general, all 'situations' are a decision problem in that every aspect (well, most anyway) of a business can be changed on a frequent basis. The animals can be fed more, less, the same; the field can be cultivated, or not; and so on. The trick is to first recognize when change, or the institution of something new, is a possibility, and then to work out what change, if any, is optimal.

This process of problem recognition depends heavily on the concept of benchmarks. These are standards held, normally in the manager's mind, that provide a comparison. Thus, if the output of standard loaves per tonne of flour is, say, 1700, then if the baker is only getting 1610, an investigation and possible action is required. On a broader scale, as another example, if each loaf is only attracting a price of US$1.58, then the marketing system may need investigating. Most managers have these benchmarks firmly held in their mind, as does everyone else in the community, but for different systems. These benchmarks must cover every aspect of the business against which judgements can then be made.

The big question is from where should the benchmarks come? A common approach is to survey businesses in a similar environment and use the production figures from the best of these businesses as the guiding benchmarks. Some would argue that this is not logical in that what is good for one business may well not be good for another, as, for example, the managers' objectives might be quite different. Similarly, the 'best' businesses might not in fact be particularly efficient anyway.

The correct set of benchmarks should, in theory, be worked out from applying production economics theory to each particular business using the manager's objectives. However, as this takes considerable work, the benchmarks from the 'best' businesses are frequently used. This approach has the added advantage that managers, like most people, are very keen to see what their neighbours are doing on the basis of being competitive. In the end, a simple and sensible approach is to rely on modified 'best' business figures where adjustments are made to suit the particular business. Thus, while the district 'best' might produce, say, an average of 1200 kg of spun wool per labour unit, the particular business might have slightly better machines and highly trained staff, producing 1300 kg, which might be appropriate when using the marginal return–marginal cost equality principle.

Dynamic Approaches

In contrast to the linear approach to decision making, many would argue that a constantly changing approach is taken by many managers. One particular study stands out in emphasizing the dynamic nature used by some managers. Ohlmer *et al.* (1998) studied a group of primary producers as they made various decisions and summarized their approach with the following diagram (Fig. 4.1).

It will be noticed that four basic phases are suggested, with each having sub-phases that the manager may use. The phases are quite broad and thus the need for sub-phases. It was suggested that the process was not linear in that the manager might duck and dive among the various phases as the situation dictated, and as the manager evolved his thinking.

Ohlmer and his co-workers discovered the managers did not necessarily work out exact figures but used a qualitative emphasis in the analysis and choice phases. They believed that when new information was obtained, the managers continually updated their:

- problem perceptions;
- ideas of options; and
- plans and expectations.

This makes common sense.

It was also pointed out that managers prefer a 'quick and simple' decision approach rather than a detailed, elaborate approach, and that they often conduct a small test before full implementation. This is also common sense where there is a lack of refutable evidence that a decision should be immediately and fully implemented.

Furthermore, managers continually check clues, as implementation occurs with a view to making improvements. Again, it is common sense to continually evaluate and adjust plans to suit the new conditions rather than waiting until the end of a cycle before evaluation in the cycle of:

- planning;
- execution;
- evaluation; and
- replanning.

Effectively, evaluation is constant and ongoing, as is information collection. It is also related to managers' networks and gatherings in that there is a constant need for information and thought. It is very useful for managers to be able to discuss ideas with others in the search for a robust system, as it is to discuss proposals with family members.

Overall, it can be concluded that a dynamic approach that uses all available information as it appears, with constant adjustments and revisiting phases in the total decision process, is the appropriate and logical way to operate. Thus, managers, to be good managers, need to constantly monitor, evaluate and replan in

Phases	Sub-processes			
	Searching and paying attention	**Planning**	**Evaluating and choosing**	**Bearing responsibility**
Problem detection	Information scanning and paying attention		Consequence evaluation. Problem?	Checking the choice
Problem detection	Information search; finding options		Consequence evaluation. Choose options to study	Checking the choice
Analysis and choice	Information search	Planning	Consequence evaluation. Choice of option	Checking the choice
Implementation	Information search. Clues to outcomes		Consequence evaluation. Choice of corrective action(s)	Bearing responsibility for final outcomes. Feed forward information

Fig. 4.1. The basic phases for dynamic approaches. (From Ohlmer *et al.*, 1998. See the Acknowledgements, p. xv, for reprint permission details.)

a completely dynamic operation. This may be uncomfortable for some in that it does require constant thinking. Some would prefer to stick to standard routines and operations. This is where the process of management is linked to management style, intelligence and a suitable personality.

Modifications to the processes

Managers take many shortcuts in making decisions compared with the simple linear, or even the dynamic, process. They have found this necessary and desirable in seeking efficiency. For example, while it is theoretically correct to ensure all possible alternatives are considered when choosing a path to follow, this seldom occurs in reality. One reason is that a quick review of the possibilities often concludes that some alternatives do not need further analysis to rule them out. An obvious example is the removal of low-value products requiring heavy working capital input from the list of possibilities when working capital is very limited and expensive; another is producing products that require an intensive production system taking many weeks when the facilities required are expensive to maintain and operate.

In addition, using the production economic decision rules allows many possibilities to be crossed off the list before formal analysis. For example, basic manufacturing machines which create a product extremely efficiently are likely to be still on the increasing output section on the production function so will not be at the optimal point (marginal return = marginal cost). In contrast, using more sophisticated, capital-intensive machines may move production into the diminishing returns area of the production function where it is more likely marginal cost will equate with the marginal return, thus increasing profit.

Another common approach is to find an acceptable solution in contrast to the optimal solution. The latter requires all alternatives to be assessed, whereas the former involves searching and evaluating systems until an acceptable one is found. This is quite common in many walks of life. For example, often when looking for a car to buy, a list of requirements is created. When a car that meets the list is found, the search is stopped well before all alternatives are evaluated.

If the list cannot be satisfied with extensive searching, the decision maker will usually revise the list downwards until it is possible to gain satisfaction. Whether this reduced search approach is acceptable will depend on the manager's objectives. In that it takes time to research out all possibilities there is a cost involved, and also a brain-power exhaustion factor. Some would judge that eventually further analysis will not provide a net positive return.

Further to this last point, all decision processes have a cost both in time and mental difficulty. The production economics rule of going to the point of equating marginal return with marginal cost applies just as much to decision analysis as it does to production decisions. However, the value of the time will depend on its opportunity cost. If the manager has spare time that is not useable in either an economic, or a leisure, sense, then there is no actual opportunity cost. However, as more time is put into decision making eventually the opportunity cost will become positive, as other productive jobs have to be put aside, thus reducing net income. In that many managers are very busy, a decision on how much effort to put into decision making is an important and real decision in its own right. Also it must be remembered that man does not live by bread alone, so other factors come into the choice of how much effort to put into analysing situations. In some cases pride will have an impact (being the best in the locality is a driving force for some), as will the obvious appearance of the business relative to neighbouring businesses.

Constructs

As part of dynamic decision making few managers actually formally analyse the day-to-day decisions. Over the years they have learnt from hard experience what works, and what does not. From this experience they develop what can be called 'rules of thumb'. These are mental instructions on the best decision given specific conditions. For each circumstance and situation a different 'rule of thumb' comes into play.

An example might be when employing labour: an applicant should not be put on the shortlist unless they have some sort of tertiary training. Such a rule might come from past experience as a general rule, but clearly there are many examples of productive people gaining their skills from experience and self-help study. The trouble is sorting out which people reliably fit this category. One exception might be if the references from past work suggest otherwise, and the manager knows and trusts the people providing the references. If this is a possibility, perhaps the manager should develop some sub-rules to apply to people without tertiary qualifications. Equally, not all tertiary trained people will be a total success.

These decision rules probably start from talking to other managers who have more experience, or from watching a business falter. As the years progress the rules will likely change based on observing outcomes and seasons. Similarly, as prices change over the years the rules may need modifying to suit the marginal cost–marginal return rule. With higher product real prices it may pay to increase the labour force of better trained people working with sophisticated, higher-costing machines.

When decision rules do not work in the sense that benchmark growth and production rules are not achieved, the successful manager will start to ask her/himself questions on what went wrong. Technically, what is known as cognitive dissonance occurs. This refers to a feeling of knowing something is wrong. Humans adjust and move to prevent cognitive dissonance.

Everyone has experienced uncomfortable situations from which decision rules are changed.

A psychology researcher (George Kelly) formalized the idea of 'rules of thumb' with his concept of 'constructs'. He believed everyone developed constructs to aid their everyday living. These constructs evolved so that a person could deal with her/his outside world and feel comfortable. People might, for example, find that being polite (a construct) was the best approach to your teacher. Hard experience might have taught this lesson. Kelly had the concept of 'man the scientist' such that 'man' experiments with how to react to situations until a 'rule of thumb' (construct) is developed that s/he believes works well for them.

Subsequently, this concept has been applied to all works of life. One example is the health field, in which people develop ideas on what works for them in maintaining health. As conditions and situations change the experiments continue, leading to changes in a person's set of constructs. Managers are no different. In some situations formal analysis might occur if conditions change sufficiently, in contrast to a mental analysis for small changes. Thus, for example, when prices change markedly relative to each other, a manager might calculate budgets to compare the economics of different products leading to a new set of rules indicating which products to produce under different relative price and cost scenarios.

Intuition (or tacit knowledge), and experts

Many managers, and people in general, make decisions without much thought, or seemingly without thought or analysis. Many an excellent manager is said to have a good intuition in that somehow they seem to inherently know the right decision and have the skills to implement the right actions and make adjustments as required. This 'intuition' is sometimes called 'tacit knowledge' in the literature. This concept has been researched to a limited extent with a view to characterizing the features of good intuition. The results suggest there is nothing particularly magical about excellent intuition.

People with these abilities probably have very good skills at learning from experience and observation, and somehow take on the correct constructs which are then stored for future use. Such people are probably reasonably intelligent and while many cannot explain the processes, they have subconsciously worked out the appropriate rules of thumb to get the job done efficiently. As time goes by, and they are exposed to the right experiences, they build up a bank of successful constructs, and they know when each should be applied.

In trying to develop good intuition it is important to practice problem-solving situations as frequently as possible, and to do so with peers and mentors enabling discussion and comparisons. Learning to ride a bicycle is probably similar. You keep trying and slowly the painful experiences lead to inbuilt systems that work; you balance, stay on and make forward progress. Gaining reasonable decision constructs is probably little different.

As intuition is such an important decision process constantly used by all managers, a separate chapter (Chapter 7) is devoted to its study. It contains a full analysis of the components which determine the success of a manager's intuition, and a discussion on how intuition might be improved.

Part of intuitive decision making may well relate to what is called 'pattern matching'. Many people work through visual images of some kind. Thus, when you see, for example, a particular cut of meat, or perhaps a type of wool carpet, you instantly know what it is without analysis. This is due to holding in memory a picture of the object so that when observed, the brain pattern matches and comes up with the solution. Before this pattern was stored in memory, trying to work out what a particular carpet's properties were required using first principles – or in other words, a formal analysis was required. This would consist of seeing the length of the staples, the level of crimping and the diameter of the wool fibres as well as their tensile strength. A textbook on the features of each wool type and their properties could be used to provide the answers.

With experience, this process is no longer required as the appropriate pattern is held in long-term memory. The same applies to good constructs. The problem is, of course, that when conditions and situations change, further analysis is required. Thus, when the animal breeders produce new wool types, and possible treatment and dyeing systems, you have to relearn the pattern which your brain instantly uses.

It is remarkable just how adept the brain is in carrying out this pattern matching, and just how many thousands of patterns the long-term memory holds, including higher-level decision situations in addition to simple problem solving such as plant identification.

People who are very good at their job are commonly called 'experts'. Experts have a mass of successful constructs stored away that they have acquired over the years. As noted in an earlier chapter (Chapter 1), research has shown that experts have the following characteristics:

- excel in mainly their own area, or domain;
- successfully perceive large meaningful patterns;
- know which factors are important;
- are fast, and solve problems quickly with little error;
- have superior short- and long-term memories;
- see and represent a problem at a deeper level than novices;
- spend a great deal of time analysing problems when they are different or new;
- have strong self-monitoring skills;
- have good observation skills; and
- relate their judgement to the relevant objectives.

Experts must also, of course, be good at recognizing when a problem exists. They are also generally quite parsimonious over the factors they observe relative to what a novice records. Generally, the learner notices and records many variables just to be sure they have taken everything that might matter into account, whereas an expert has worked out which variables are important and so simplifies the problem to its essential elements.

Possibly the most important aspect of an expert is their self-monitoring skill. Many people have a rich set of experiences yet are not experts. Learning from each situation requires an ability to stand back, analyse the problem and gather in the essential lessons. This all relates to self-monitoring skills and a realistic view of the production processes. Effectively, it is the ability to learn from experience that is important. Few people can learn from a textbook and then be an expert, unless it is something like pure maths! Thus, the right kind of aptitude and managerial style are both important, as is education.

This discussion reinforces the relationships talked about in the chapter on the origins of managerial ability (Chapter 3) and gives the background to some of the reasons why the personal attributes are important in good management.

Both of the case study managers highlighted here believe they have developed good intuitive skills, though Margrave believes you have to be careful because conditions change and so the intuitive answer can be misleading. Hank used to walk a lot and mull over the problems and decisions to be made, and he commented 'suddenly the answer was there'. In all probability, all the thinking, together with his subconscious activity, eventually produced an answer that took account of the various factors impinging, and which resulted in his 'self' being comfortable. People keep mulling if the suggested decision just does not feel right. Managers who have the ability to get this right go on to be successful, whereas those who find 'comfort' before properly allowing for all factors end up with doubtful decisions.

Hank finds writing everything down also focuses his mind, particularly in recent times as there is less time for walking. Job lists are always a starting point to further refine decisions, which then leads to decision and action. Hank recounts the ability of his intuition over a weather event, 'We had a terrible spring and my feeling was that the autumn would be perfect, and sure enough it was and the decisions I made anticipating this situation paid off perfectly.' But, Hank points out, a bad spring is not a reason for a good autumn, it is just that in this case, 'I somehow picked up all the right signs without realizing it.'

Innovations and their adoption

New technology is constantly confronting managers, who must decide whether it is useful. While this decision is little different from any other, it is useful to list the aspects of new

Fig. 4.2. Trying out, and perhaps adopting, new ventures is always an important part of keeping ahead in the profit stakes. (Courtesy of David Hollander.)

ideas that managers tend to consider. In the end, of course, they need to decide whether an innovation will enhance their operation. Extension people work hard at introducing new approaches but are often frustrated at the speed of uptake. Usually, however, the slow uptake is for good reason. If managers can see significant benefit, they will proceed with speed (as many have in domesticating deer and the various products that the industry leads to; see Fig. 4.2).

With respect to its adoption, the aspects of a new technology that many managers think should be considered are:

- trialability;
- complexity;
- comparability; and
- observability.

If it is easy to trial the innovation without a major commitment, it is easier for the manager to come to a conclusion. Furthermore, the analysis will be much easier, as will the trials, if the innovation is not complex. Then having alternatives that are comparable provides benchmarks against which the innovation can be compared; and similarly, if the results of trials are easily observed, a decision is much easier. Where the capital requirements of an innovation are not substantial, that is also a positive aspect to any change.

Where all these factors are favourable, uptake of a worthwhile innovation will occur more rapidly than would otherwise be the case. This assumes the manager believes the innovation will be of benefit to his objectives, though the 'comparability' aspect compares it with the existing, or other, alternatives.

What a manager actually does when confronted with a new technology, relative to what he should do in a simple rational sense, can be two different things. To study this process a theory known as the 'theory of planned behaviour' (Ajzen, 1991) is sometimes used. It is hypothesized that an intention is dependent on three factors:

- behavioural beliefs;
- normative beliefs (social norm); and
- control beliefs.

The behavioural beliefs represent an individual's beliefs over the consequences of an action, its outcomes and benefits, whereas the normative beliefs are the person's beliefs about what others expect him to do. This might be the influence of a respected peer, or perhaps a spouse. In contrast, the control beliefs are the restrictions on what might be possible given the resources a manager has, and his skill set.

Using standard questionnaires, values can be given to each component leading to an assessment of the decision maker's intention which then leads to action. This theory has been applied to several agricultural situations (as well as many other industries and situations). Two examples are given by Beedell and Rehman (2000) and Sambodo and Nuthall (2009). The first looks at conservation practices, and the latter at technology adoption in less-developed agriculture in which it was discovered that another important factor in intention and action is what was called a 'bargaining process'. Effectively, family, and other, bargaining took place in order to come to a conclusion on action. This conclusion further reinforces the place of family in decision making, which is discussed fully in Chapter 9.

Other examples include Rah *et al.* (2004), who considered why women purchase soy products; and Moon (2021), who investigated the patronage of green restaurants based on the theory of planned behaviour.

Overall, the theory has proved to be very useful in assessing the factors that influence decisions and their quantification.

Objectives and Their Impact on the Managerial Processes

Maximization

Most production economics texts would have us believe that the manager tries to maximize her or his objectives through the correct choice of products, inputs and production processes. No doubt many managers would voice this objective, but in reality it is unlikely that their actions lead to maximization. The texts also frequently assume the main objective is the maximization of business profit, where this is defined as the sum of the sustainable net cash profit and the increase, hopefully, in real net asset value. Indeed, the decision rules of equating marginal return to marginal cost assume this profit goal. In reality, as strongly noted in Chapter 3, a manager and his family tend to have a more complex goal set which includes factors like leisure time, pleasantness of working conditions, products that the manager enjoys working with, protection of the environment, spending time with colleagues and friends, and so on. This is a multi-goal decision environment in which a manager can seldom succinctly voice just what the business's goals are and the trade-off rates when playing one sub-goal off against another.

Satisficing

In contrast to maximization, many managers look for a satisfactory, as against the highest, level of an objective. They keep trying to improve, say, cash income, until a particular level is reached. Searching for improvements then tends to stop. Often there is a priority listing between goals, so a satisfactory level of income becomes the first goal; and once it is reached, then, say, leisure time is addressed with a view to achieving at least a particular level of spare time; and so on down the list of goals. The ultimate situation is when all goals are met at the required level. This decision process involves maintaining adequate levels of higher-order goals as lower-priority ones are addressed. Provided the minimum income and leisure time are achieved, then maybe the appearance of the business is upgraded to meet the desired level. But this only happens if the higher-priority goals are met and maintained.

Between the maximization and satisficing approaches are mixed arrangements. For example, a manager might wish to maximize income subject to a minimum total leisure time. Thus, plans are made for this minimum leisure time, and provided this is maintained the search goes on for production and systems that maximize profit.

Multiple goals

Where several factors are in the mix making up the objective, the manager may relate the value of each to create a trade-off situation. For example, it might be considered that US$100 is equivalent to 5 hours of leisure time. If the value of leisure goes up, it is worth substituting real production for leisure time where the product price stays constant. The opposite can occur if the price of the product increases, or its production efficiency so that its net return increases. These substitution rates will vary as the quantities of products produced change. Thus, if the profits are such that leisure is sacrificed for greater profit, the value of leisure will increase as the quantity taken decreases until eventually, to the manager and his family, it is no longer worth substituting leisure for profit, and a new equilibrium is reached.

A manager might have a value in mind for each 'product' that might come from the business (leisure, profit, enjoyment of particular products, enhancing the environment, etc.) so the quantity of each is shuffled round until the marginal value of each is equated; this is another production economics principle.

As most managers do have multiple goals this mixing and matching undoubtedly occurs, so that for any particular manager there will be a mix that s/he aims for. Whether the manager tries to maximize some sort of overall satisfaction, or simply tries to reach a satisficing level of each, will depend on the manager's personality and, where appropriate, family objectives. In this process, a knowledge of what must be given up as another goal is increased is crucial to success in coming to an appropriate solution. In some cases, of course, where there are surplus resources, it may be possible to increase one goal without decreasing another. Land, or perhaps working capital, for example, might be used with more intensity to increase income without altering leisure time. This is almost a win–win situation.

Integration

Understanding that the goals held will impact on the decision processes used is important in assessing a manager's modus operandi and, therefore, how s/he might improve decision skills. Maximization is more difficult than satisficing, or some variant of it. It requires ensuring absolutely all possibilities are considered and demands much more time spent on data searching. Satisficing is more about searching around for a system that meets the basic requirements.

Whatever the manager's approach, the concept of utility applies. In theory, each output from a business gives rise to satisfaction, or utility as it is called in the production economics textbooks. Usually the level of utility from production is a decreasing function, so, for example, each additional income unit, say, US$500, provides lower and lower marginal utility. The difficulty with this theory, however, is that utility does not have a comparable and definable unit. This is in contrast to, say, length, for there is a standard metre against which all lineal lengths can be measured and compared. So, while in theory everyone has some kind of internal measure of utility, it cannot be compared across people.

In a multiple-goal situation each individual will have a utility value for each output, and this value will decline as more and more is produced. Conceptually, there is an equation that measures the utility from a particular business system. Thus, for some managers the equation might be:

$$\text{Utility} = 2Y + 4L + 1.2YL - 0.2YY - 0.1LL$$

where Y is dollar income and L is hours of leisure (note that this function has decreasing marginal utility due to the negative terms).

Given an equation like this, together with a knowledge of the relationships between the physical production and the output of income and leisure, it is possible to find the production system which maximizes utility, or in a satisficing situation, meets the minimum requirements at least cost.

In reality, a manager certainly does not sit down and work out his utility. What might happen, however, is that something like this process occurs in an intuitive subconscious operation (e.g. he might well know water sports are great fun; see Fig. 4.3). Somehow the manager allows for the substitution rates and comes up with an answer. In looking at ability, just how well this happens will help define how good a manager is, and where improvements are possible.

Fig. 4.3. Leisure and relaxation are an integral part of life and must be catered for.

Complications

So far no mention has been made of the vagaries of production. Output is uncertain (e.g. a harvesting contractor's total area harvested is very dependent on the weather, among other variables), as are the prices received, so it is impossible to predict exactly what the output of income and leisure might be, and similarly for the other outputs that might have utility. This means further concepts must be added to any analysis.

It is clear achieving goals is often destroyed by uncertainty and risk. Accordingly, for example, a manager might prefer a production system that can guarantee a certain minimum income. Or it could be that he will be quite happy to take the good with the bad if, on average, over several years the average profit is as high as possible. Generally, a manager's modification to his objectives under the realistic risk and uncertainty situation is dependent on his attitude to variability. This might be a risk-preference attitude, or in contrast, a risk-aversion attitude, or something in between.

The section 'Attitude to Risky Situations' in Chapter 2 (p. 21) contains a more complete discussion on managers' attitudes to risk and uncertainty as well as providing a question set that can be used to assess a manager's attitude.

A risk averter is happy to give up average income in return for a relatively stable income. Such a person is probably content to take out considerable insurance, for example, in that a sure cost to maintain a reasonable income is better than a higher average income that in fact might be quite low in some years. Indeed, if a manager's fixed debt servicing is quite high, a very poor year might bankrupt the business even though the business system used is, on average, the best possible.

In contrast, a risk preferer is quite happy to take the chance of a bad year if, on average, his income is much higher than with the safe-and-sure approach. Each manager will be slightly different in their attitude, and these differences probably stem from their personality and background. Also making an impact is the manager's debt situation so if, for example, debt is low and the manager has a good asset base, a manager is more likely to take risks.

Sorting out a manager's risk attitude is a precursor to understanding how efficient each one is. Their attitude affects what is a correct choice, and this will be different for each manager. Thus, one manager might be rational in diversifying production, whereas for another diversification would be inefficient. Diversifying probably means producing a product of lower profit, but with the hope that in a poor year for one product, one of the others will be having a good year. Thus, judging managerial ability and the processes used by a manager does depend on the manager's objectives and his attitude to risk.

Many books have been written about optimal decisions under risk, and how to formally assess a manager's attitude, and to calculate risk levels. It is not appropriate to repeat these procedures here, but a reader interested in this detail should consider, for example, Anderson *et al.* (1977) and Verbano and Venturini (2013) as two examples. Furthermore, some workers believe that in many situations it is difficult to quantify the chances, or probabilities, of outcomes and so a body of theory about decisions under uncertainty has been developed. (Often the word 'risk' is used when talking about situations where a chance, or probability, can be placed on each outcome. In contrast, where you know a range of possible outcomes is possible due to chance, but you cannot say whether one is more likely than others, the word 'uncertainty' is used. Tossing a coin or die, for example, is a risk situation, whereas the price of genetically engineered cotton is most likely an uncertainty situation as there are few records to guide probability estimates.)

At the heart of these uncertainty situations is 'game theory' (Agrawal and Heady, 1972), in which the manager is regarded as making decisions with nature opposing them. One of the proposed decision processes is for the manager to consider each strategy and work out the likely income from each under representative states of nature. Then a choice is made by selecting the strategy that has the highest of the lowest payouts possible from each alternative; this is the maximum–minimum rule. Various other decision rules are possible too. One is what is known as regret theory. Here the pay-offs are listed as the regret the manager would experience if a particular state of nature occurred for each strategy. With this data, a manager can then select a strategy that minimizes the regret.

There is no doubt that classifying and quantifying managers' objectives is a difficult area, as the range of possibilities mentioned demonstrates. For each manager the reality will be some mixture of the ideas covered. Anyone trying to understand how a manager operates needs to take account of the manager's objectives and relate them to how the manager might be made more managerially efficient given his objectives. A manager, for example, very concerned about a stable income must be able to assess probabilities and how they can be used in decision making. Without these skills, whether held intuitively, or formally, he will not be a good manager. In contrast, someone who is risk-neutral need not be as skilled in risk and uncertainty analysis as these aspects are less important in their objectives.

Similarly, where a manager has multiple goals this must be recognized and allowed for in decision making, whereas someone focused on simply maximizing average net profit needs less understanding about goal substitution and their relationships in order to be a good manager. Effectively, 'goals and objectives' is a further complicating factor in judging and training managerial ability.

Concluding Comments

All managers are unique in terms of their objectives and goals, and in their management style and aptitude. Consequently the decision processes used will be different. In improving managerial skill it is important to understand the processes actually used, and their relationship with the manager's objectives, if improvements are to be made.

Given that production is almost always undertaken under risk and uncertainty, a successful process is likely to be dynamic as the conditions and outcomes are always changing, and frequently in an unexpected way. Thus, the process must ensure observations are constantly being updated, and that the manager looks well into the future when making plans. Decisions made now will be played out into varying future time spans. To make the right decision it is important to know its range of possible long-term ramifications to ensure a complete appraisal. And in making the decision, the manager must clearly use the logical set of decision rules which have been well enunciated in production economic texts.

Managers must, of course, be ahead of the play by recognizing when problems and opportunities arise. It must be recognized that some managers do simply 'follow the leader' in contrast to carrying out their own analyses. If the leader is successful, and their situation is virtually the same as the case in point, the manager is lucky – especially if his implementation skills are up to the task. This is the exception, so managers must be aware of the outcomes and efficiencies that are possible if they are to be good managers. 'What is possible' gives rise to the benchmarks.

Implementation of the decision processes must recognize the manager's specific objective set, and appropriate modifications to standard procedures put into effect. Especially if the manager is other than risk-neutral, due allowance must be included in analyses that take variability into account. And where there are multiple goals, calculations, whether formal or mental, must allow for each objective, thus complicating the process compared with simple profit maximization.

In working to improve the processes that make up managerial ability, it is important for managers to discuss the issues with mentors of various kinds. These might include family members, but also manager friends. It is through constructive criticism from self and colleagues that processes are improved. The objective is to become an expert with the associated attributes. Practice, and discussion, helps create the skills necessary. Success creates rewards which reinforces the developing process, and vice versa. Eventually, the result is a permanent change in behaviour.

In that goals tend to be dynamic, with maturity and experience people change what they want from their business life, so an optimal decision process is also dynamic. Sometimes feelings change from day to day, so care must be taken in sorting out a robust process that serves well over time. Furthermore, when it is clear that some goals are impossible to achieve, people rationally lower their sights somewhat to create practical aims and outcomes.

In a nutshell, success depends on:

- recognizing problems;
- making the correct observations;
- successfully applying the correct decision rules;
- successfully implementing the decisions made; and
- successfully monitoring and adapting the decisions in response to the ever-changing situation resulting from risk and uncertainty.

5 Skills Required

Abstract

A good manager will clearly have a range of skills depending, in part, on the type of business. This chapter contains reviews of opinions on the main skills required, including an ability to observe the relevant factors, risk management skills and the like. Results from two surveys asking managers and, secondly, business consultants to rate the important skills are presented and discussed. These covered the general areas of managerial attributes, entrepreneurial skills and personal attributes. The summary of the results covers the important skills (e.g. making requirements clearly understood, identifying key factors, meeting deadlines, marketing, early observation ...). Further thoughts on other essential skills are also introduced. Important examples include risk management, listening skills and key issues observation, visualization and self-observation including self-critiquing. The importance of the correct analytical skills is also emphasized. The chapter ends with the views of the case study people, one of whom stressed 'the business must be enjoyable'.

Introduction

Any manager must have certain skills as a minimum, but for excellence the manager must be good at the complete list of competencies, or skills, necessary to operate a business. There are various opinions on the skills that make up the list, and no doubt this varies to a certain extent with the type of business, but there is certainly a core set common to all situations. You can probably quickly come up with a list of what is necessary after some thought. A successful manager must have, first,

- good technical skills as without them the jobs will not be successfully completed, particularly if the manager her/himself performs physical work.

  Similarly,

- a good knowledge of the technology of the business involved is essential.

For example, understanding how the market might respond to changing a marketing plan is essential for correct decisions. There is a myriad of such technical 'facts, figures and relationships' that are a prerequisite to good decision making, as is the ability to apply the decisions.

Other examples of the skills required include:

- observation ability;
- visualization, or prediction, of likely outcomes resulting from any decision;
- negotiation skills;
- people relationship and management skills;
- risk management skills;
- an ability to simplify situations down to their essential elements; and so on.

This chapter contains an outline of just what these generic skills are likely to be. This is achieved by reviewing information from a survey seeking managers' views of what skills are important, and then assessing their views, and adding to them as required.

While having an appropriate aptitude, personality and experience are all essential, in a specific sense all these attributes must be brought to bear on each of the skills required. In the end, it is the level of the skills that determines the outcomes, with the basic building blocks being the precursors to successful training in each of the skill areas. Training, of course, is often an 'on-the-job' process in contrast to formal coursework.

This chapter proceeds with the results of the manager survey, and then a discussion on the skill sets. Some of the data presented and discussion is based on Nuthall (2006). The chapter finishes with an introduction to two professional consultants and their views on the required skills for managers. The views of these

DOI: 10.1079/9781800627246.0005

two consultants on training methods, and on ways managers can improve their skills, particularly with respect to the place of consultants, are presented in Chapter 10.

The Questionnaire, Sample and the Respondents

The questionnaire was developed from the literature on competencies for a range of countries, and by discussing the possibilities with managers and consultants. The questionnaire used is presented in Appendix 5A. The potential competencies were broken down into three groups:

- managerial attributes;
- entrepreneurial skills; and
- personal attributes.

This division ensured the set to be scored was kept to manageable proportions. The distinctions were somewhat arbitrary.

The random sample of 2000 managers (which was approximately 5% of the total population) was divided into:

- 16 statistical regions;
- six business-type groupings (intensive and extensive, and different food and fibre product types); and
- 12 business-size groups.

The number selected from each group was based on the proportion of the total population in the group; 823 useable responses were obtained giving a response rate of approximately 41%.

The managers responding to the sample, on average, had a reasonable formal education. Some 2.3% had no more than primary education, but 35.7% had up to 3 years of secondary education with another 28.6% completing more than 3 years. As for tertiary education, 13.6% of the sample attended formal qualifications requiring up to 2 years, with the remainder, 19.5%, having more than 2 years of tertiary education.

In terms of business size it is estimated that in 2022 values the total asset value per business was approximately US$5,464,800 with an average debt of $1,352,400, leaving net assets at $4,492,400.

Managerial Attributes

Respondents were asked to rate the importance of a range of attributes on a 1 ('not at all important') to 7 ('very important') scale. Table 5.1 gives the results for both the managers and a sample of professional agricultural consultants (approximately 320).

An analysis of variance showed the differences between the means were highly significant with p=0.0 (F = 199.05 and 236.45 for consultants).

While the list is ordered according to the managers' ranking, the ranking according to the consultants is given in parentheses. The order changes slightly, but generally the two groups agree on what is important.

The three most important attributes were:

- observation;
- introspection (key factors and priorities); and
- communication.

These embody the four highest-ranked attributes for both groups of observers.

Eight of the 15 items are scored 5.5 or better, indicating many attributes are considered important. This might be expected as only the most likely were included in the list offered. To help analyse the responses a factor analysis was carried out. This looks at the correlations between all the items to isolate the groups that tend to go 'hand in hand'. Studying the components of each group might well suggest some basic attributes that underlie those listed. Table 5.2 contains the results of the factor analysis.

There were two underlying factors which explained 45% of the total variance. Note that factor loading values less than 0.3 have not been presented as they contribute in only a minor way to the factors. The data is interpreted through noting that, for example, Factor One is made up of:

- 72% (or 0.72) of item 9 (making requirements clearly understood);
- 68% of items 1 and 15 (identifying key factors and assessing job priorities);
- 62 and 61% respectively of items 11 and 2 (knowing how to choose between

Table 5.1. Importance of managerial attributes. Mean scores on a 1–7 scale (ranging from 'not important' to 'very important'). (From Nuthall, 2006. See the Acknowledgements, p. xv, for reprint permission details.)

		Managers	Consultants	(Order)
1.	Being up to date with the current condition of the business in its totality (bank balances, animal condition, crop growth, soil moisture, feed levels, machinery repair)	6.23	6.07	(3)
2.	Ability to identify the key factors in a problem, and discard the irrelevant	6.16	6.29	(1)
3.	Making requirements clearly understood (effective communication)	6.13	6.28	(2)
4.	Assessing job priorities	5.93	6.00	(4)
5.	Quickly analysing and sorting out situations that have never been faced before	5.68	5.26	(12)
6.	Having a clear understanding of the family's objectives, values and goals, thus making assessing the value of alternative actions easy	5.67	5.79	(5)
7.	Picturing (understanding) the consequences of a decision over the many (or few) months/years it might impact over (e.g. planting an area in forestry, subdividing a building)	5.63	5.71	(7)
8.	Being able to efficiently organize and carry out quite complex operations (e.g. get a new packing shed operational on time)	5.61	5.52	(8)
9.	Developing appropriate and detailed plans for both short- and longer-term horizons	5.47	5.71	(6)
10.	Understanding the basis on which to choose between alternatives (e.g. knowing how to cost unpriced labour, knowing how to do gross margins, understanding diversification principles)	5.31	5.32	(11)
11.	Skill at keeping, interpreting and using recorded data about the business and associated factors (e.g. market trends)	5.17	5.42	(10)
12.	The ability to predict product prices into the foreseeable future, or at least understanding the factors that determine the prices, and understand market requirements	5.16	4.96	(13)
13.	Developing and maintaining a support network of colleagues and professionals	4.89	5.44	(9)
14.	Being able to predict local weather better than the official forecaster	4.23	3.07	(15)
15.	Understanding the local political scene as it might impact on rules affecting what can be done	3.88	3.40	(14)

alternatives and quickly sorting out new situations);

- and so on for the rest of the items.

In Factor Two the most important item is:

- at 80%, 'understanding the local political scene'.

However, this item has a low ranking (15th); while it is a major contributor to the factor, this does not necessarily mean the factor as a whole is important. Indeed, if the scores of each factor component are averaged, Factor One is rated 5.62 and Factor Two is 4.99, indicating the importance of Factor One.

Table 5.2. Factor analysis loadings (item contributions) for the managerial attribute variables (refer to the questionnaire list in Appendix 5A for the attribute represented by each number).

Attribute number	Factor number ONE	TWO
1.	0.68	
2.	0.61	
3.	0.42	0.31
4.		0.66
5.		0.80
6.	0.30	0.53
7.	0.47	0.38
8.	0.56	0.38
9.	0.72	
10.	0.54	0.45
11.	0.62	
12.	0.58	
13.	0.48	0.51
14.	0.33	0.62
15.	0.68	

While the managers believe there are two basic sets of attributes that go together, the consultants believed there were four groupings (with average scores of 5.66, 5.46, 5.84 and 3.97). Clearly the consultants believed managerial skill was made up of more factor components than the managers. However, the items left out of Factor One by the managers were also left out by the consultants. Given that the score rankings were much the same for both groups, the conclusion must be that the most important attributes have been clearly stated.

The above analysis has grouped all managers together no matter their business type or personal attributes. Consequently, the analysis was repeated for a wide range of different groupings including:

- business type;
- age;
- gender;
- managerial style;
- education;
- self-assessed intelligence;
- managerial ability;
- objectives; and
- whether a business computer is used.

Generally it was concluded that the results were largely the same no matter which groupings were compared. Managers, therefore, had the same views no matter which business type, location or characteristic set they belonged to.

Entrepreneurial Skills

Respondents were asked to rank 12 entrepreneurial skills on a 7 ('very important') to 1 ('not at all important') scale. They were also given the opportunity to add further skills if they thought those offered did not cover the full list of possibilities. While some respondents did write in their thoughts on additional skills, most were a rewording of the 12 listed skills. Table 5.3 lists the scores given on a 1 ('not important') to 7 ('very important') scale ranked in order of descending importance. The scores given by the consultants are also presented.

The F significance test showed the differences in the means were highly significant (p=0.0, F=49.48). The five most important skills involve:

- meeting deadlines;
- successfully obtaining decision information;
- price negotiation;
- successfully handling risk; and
- an anticipatory intuition.

These skills are scored very similarly except for the top priority of meeting deadlines – this stands out.

The consultants have a similar priority list so there is general agreement, though the 'ability to learn new skills' is ranked lower by the managers, as is a belief in being able to control what happens. Clearly consultants believe managers still have some learning to do!

As for the managerial attributes, the managers believe a wide range of skills are important. Nine of the 12 listed skills are ranked 5.55 or greater. The lowest ranked, by both the managers and consultants, was the ability to forecast longer-term opportunities – perhaps they do not believe there will be new

Table 5.3. Importance of entrepreneurial skills. Mean scores on a 1 ('not important') to 7 ('very important') scale.

		Managers	Consultants	(Order)
1.	Understanding deadlines and being able to 'act in time' (e.g. spray before insect damage, fertilizer applied in good time)	6.16	6.38	(1)
2.	An ability and determination to look/ask/seek out information thought to be necessary for making decisions	5.78	5.99	(2)
3.	The skill to negotiate the best possible deal (price, arrangement)	5.78	5.34	(9)
4.	Understanding sources of risk and what can be done to reduce its impact	5.75	5.70	(4)
5.	An intuition that gives early-warning signs when something is not right, or, in contrast, when something positive needs exploiting	5.75	5.65	(6)
6.	Ability in learning new skills	5.58	5.71	(3)
7.	An ability to look ahead and anticipate likely problems, needs and opportunities	5.70	5.61	(7)
8.	When faced with opportunities, ensuring that ALL alternatives are sought out, considered and evaluated	5.65	5.35	(8)
9.	A belief in being able to control a lot of what happens around the business; in contrast to a belief that not much is really controllable due to the weather, markets or government action	5.55	5.69	(5)
10.	Skills in finding the very best market (price, quantity, etc.) for all output	5.34	5.03	(11)
11.	Being able to seek out, identify and clarify new opportunities (production, products, marketing, etc.)	5.21	5.25	(10)
12.	The skill and intuition to forecast well into the future likely opportunities in products and production systems	4.90	4.68	(12)

and promising opportunities and that improving on existing systems and products is more relevant.

To examine whether managers believe there are inherent groupings among the skills a correlation (factor) analysis was conducted. Table 5.4 presents the results (loadings less than 0.3 are not presented due to their insignificance).

Compared with the consultants, the managers saw entrepreneurial skills more simply in that they grouped them into two factors (that explain 56% of the variance) instead of three (explaining 59% of the variance). Interestingly, the important skills in Factor One are:

- information seeking (0.75);
- an ability to learn new skills (0.75);
- dealing with risk (0.72);
- an ability to look ahead (0.66);
- a full comparison of alternatives (0.64);
- early-warning sign intuition (0.63); and
- belief in being able to control many factors (0.61).

Table 5.4. Factor analysis loadings (item contributions) for the entrepreneurial skill variables (refer to the questionnaire list in Appendix 5A for the skills represented by each number). (From Nuthall, 2006. See the Acknowledgements, p. xv, for reprint permission details.)

	Factor number	
Skill number	ONE	TWO
1.	0.35	0.62
2.	0.75	
3.	0.75	
4.	0.63	0.32
5.		0.88
6.	0.50	0.39
7.		0.76
8.	0.61	
9.	0.64	
10.	0.39	0.68
11.	0.66	0.35
12.	0.72	0.32

All these skills are seen as a connected 'kit bag' involving common-sense data collection and analysis, and a perceptive observation system that is tuned to opportunities and problems.

The important components of the second factor are skills in:

- marketing (0.88);
- negotiating (0.76);
- forecasting intuition (0.68); and
- an ability to discover new opportunities (0.62).

These are clearly connected. Which factor is more important? Factor One has an average score of 5.60, and for Factor Two it is 5.57, so both must be regarded as significant.

With respect to subgroupings based on variables such as business type, while there were some statistically significant differences between some of the mean scores for the different groups, the rankings of the various skills changed only marginally.

Personal Attributes

The questionnaire included 18 personal attributes that the respondents were asked to score on a 1 ('not important') to 7 ('very important') scale. Table 5.5 lists the attributes in score order. An analysis of variance indicated the differences in the mean scores were highly significant (p=0.0, F = 207.20).

It will be noted from Table 5.5 that seven of the attributes have a score greater than 6.0, and another three are greater than 5.7. Overall, the respondents have scored the personal attributes rather higher than for the managerial attributes and entrepreneurial skills. Clearly, early observation of important indicators is regarded as the top attribute or skill, and the ability to learn from experience and developing a 'good moral character' are not far behind.

Maintaining good relationships with business associates outside the business and 'keeping cool' under all circumstances are also scored among the top attributes. Being able to act decisively and good relationships with employees or contractors are also highly rated. At the other extreme, high intelligence and good school grades, and developing a strong personality, are not regarded as being particularly important. Similarly for a tertiary education in areas related to primary production. Yet, it must be noted that approximately one-third of the respondents had experienced tertiary education, though the areas of study are not known.

With respect to the attribute rankings, it is interesting to note that at least in the area of personal attributes the consultants are virtually in agreement with the managers, both in the ranking and some of the score levels (highest is greater, but lowest is lower).

To assess the groupings of the attributes, a further correlation (factor) analysis was conducted. The loadings for values greater than 0.3 are given in Table 5.6. The three-factor solution (with eigenvalues ≥ 1) explains 53% of the variance.

Factor One is all about personality and relationships:

Table 5.5. Importance of personal attributes. Mean scores on a 1 ('not important') to 7 ('very important') scale. (From Nuthall, 2006. See the Acknowledgements, p. xv, for reprint permission details.)

		Managers	Consultants	(Order)
1.	Early observation of important indicators around the business (e.g. lambs are scouring, wheat is infected, cows losing weight, pasture growth has increased, etc.)	6.65	6.72	(1)
2.	Ability to learn from experience, mistakes and failures	6.35	6.28	(2)
3.	Developing a 'good moral character' involving openness, integrity, reliability, trustworthiness, etc.	6.35	6.10	(3)
4.	Maintaining good relationships with outside people – bankers, accountants, suppliers, etc.	6.19	5.87	(6)
5.	Keeping a cool head and putting aside any tendency to panic when faced with stressful situations	6.19	5.79	(7)
6.	Having the confidence to draw conclusions and act quickly and decisively	6.18	5.95	(4)
7.	Obtaining employees' and/or contractors' cooperation and understanding leading to harmonious and productive relationships	6.08	5.91	(5)
8.	Understanding the interrelationships between all the components of the business (e.g. rainfall–soil moisture–plant growth–animal grazing, i.e. what affects what?)	5.99	5.77	(8)
9.	Successfully resolving conflicts on, and off, the business (e.g. dispute between employees)	5.78	5.57	(10)
10.	Successfully judging personality and selecting suitable employees	5.74	5.53	(11)
11.	An excellent knowledge of facts, figures, procedures and methods, with respect to soils, plants, animals, machines, buildings, etc.	5.58	4.99	(12)
12.	Accepting the good and the bad and not letting it affect management and decision making	5.53	4.93	(13)
13.	High motivation in constantly seeking better ways and implementing them; in contrast to being happy with current systems	5.28	5.75	(9)
14.	The determination to keep working all hours, until the high-priority jobs are completed	5.24	4.48	(15)
15.	Being prepared to give it a go and take risks in changing production systems and/or starting new ventures	5.14	4.84	(14)
16.	Developing a strong personality so that others 'sit up, notice, respect and act' on what is said	4.96	4.27	(17)
17.	Tertiary education in areas related to business production (marketing, production economics, etc.)	4.61	4.33	(16)
18.	Having above-average intelligence and school grades	4.46	4.19	(18)

Table 5.6. Factor analysis loadings (item contributions) for the personal attributes (refer to the questionnaire list in Appendix 5A for the attributes represented by each number). (From Nuthall, 2006. See the Acknowledgements, p. xv, for reprint permission details.)

	Factor number		
Attribute number	ONE	TWO	THREE
1.	0.50	0.50	
2.	0.54	0.50	
3.	0.58	0.51	
4.		0.46	
5.		0.70	0.32
6.		0.71	0.34
7.	0.34	0.54	
8.	0.58	0.48	
9.		0.38	0.33
10.	0.68		
11.			0.60
12.	0.49	0.45	
13.	0.70		
14.			0.69
15.			0.76
16.	0.73		0.32
17.	0.75		
18.	0.70		

- developing and maintaining good working relationships both within and outside the business;
- the ability to learn from experience;
- early observation;
- a personality that does not panic; and
- acting quickly when required.

Perhaps 'early observation' is out of place as a cohort, but possibly it is related to a personality that is careful and gives attention to detail.

Factor Two has as its important components what might be called:

- an adventurous spirit ('give it a go', take risks, keenness to try new ways, and so on); as well as
- early observation;
- acting decisively; and
- not panicking.

The synergies here are clear.

The important components of Factor Three are above-average intelligence, tertiary education and a strong personality. All these factors are not regarded as being relatively important, but they clearly relate to each other, or at least the first two do. The average score of the components for each factor is 6.09, 5.81 and 5.06, respectively. The first two factors dominate in importance.

As before, the rankings remain relatively stable when the full ranges of subgroupings are compared.

Bringing Together the Managers' Views on Competencies

The data as presented does not provide very distinctive and clear-cut conclusions on a shortlist of the most important competencies. A simple list of, say, six to ten competencies with scores well above the rest did not emerge. The respondents believe there is a wide range of skills that are part and parcel of managerial success, and any line used to divide an important and less important group will be at an arbitrary position. It was therefore important to look for correlations between the higher-ranked competencies to see if the members of the groups have similarities that can be used as core factors in training programmes – thus the various factor analyses that were presented. To further enhance the factor groups, the competencies from all categories with a score greater than 5.69 were grouped and reanalysed. Table 5.7 gives the results. Using 5.69 as the divider provided 18 items with a factor loading greater than 0.5.

The factors explained 54% of the variance with a very high level of significance (p=0.0).

It is clear Factor Three is about good skills in selecting and managing people. Factors One and Two are more complicated and involve several competencies. Factor One is about planning and associated issues such as information gathering and risk management. It is also about effective implementation of the plans:

Table 5.7. Factor analysis loadings (item contributions) for competencies from all groups with a score greater than 5.69[a].

	Factor number		
Competency (paraphrase)	ONE	TWO	THREE
Observing current state of business		0.57	
Planning for short and long terms	0.52		
Obtaining planning information	0.59		
Intuitively noting early signs	0.64		
Acting on time	0.65		
Negotiation skills	0.65		
Looking ahead and anticipating	0.71		
Good risk management	0.73		
Early observation of important factors		0.69	
Keeping a cool head		0.66	
Confidence to conclude and act		0.62	
Learning from experience		0.63	
Developing a good character			0.57
Understanding interrelationships		0.56	
Getting cooperation of employees/contractors			0.59
Successful judge of personality			0.77
Resolving conflicts			0.80
Good relationships off the business			0.64

[a]Only loadings of 0.5 or greater are displayed.

- looking ahead and anticipation;
- intuitively picking up important signs;
- successful negotiations; and
- acting on time.

Effectively Factor One is about planning, implementing and control. Factor Two is similar and really reinforces the implementation component of Factor One through:

- early observation skills;
- keeping a cool head;
- confidence to decide and act quickly;
- learning from experience; and
- understanding all the interrelationships between the components of any system.

These factor analyses make the managers' views of the components of good management very clear.

Furthermore, these components are relatively stable across different SME business types as it was shown the rankings change very little with age, education, business type, managerial style, gender, profit objective variations and computer ownership.

Summary of Skills Required

Introduction

Other than the skills rated by the managers, some thoughts suggest further skills likely to be important in good management. As these slot into the rated list, the managers' list will be summarized with these additional skills introduced. These are not presented in any kind of priority order for they are all important core skills,

Fig. 5.1. Diversification can be an important way to reduce risk. Managers must use their imagination to come up with solutions – in this case, developing a large honey production and processing business.

though in each business situation some will be more significant than others.

Risk management

First, RISK MANAGEMENT is recognized by the managers as being important. Risk and uncertainty are certainly a dominating factor within rural SMEs in most countries due to the weather and markets in particular. Every manager must be aware of the nature of the risks, and what can be done about alleviating the problems. Just how much alleviation will clearly depend on the manager's objectives and attitude to risk as previously discussed. Selecting the right alleviation systems to suit the business and objectives is, therefore, an important skill. This means being fully aware of all the strategies such as diversification (which can take many forms. See Fig. 5.1), selection of low-variability production processes (e.g. good and supported supply chains) and products, use of contracts and forward selling, insurance, flexible systems and products that can be changed to suit the situation, and so on.

Observation

Second, and perhaps before risk management, comes OBSERVATION as noted by the managers. They specifically talked about noting the early signs of relevant situations so that action can take place in good time, and about acquiring accurate planning information through observation both on and off the business. This observation must also encompass problem recognition.

Not talked about by the managers was the skill in knowing what to observe. Many managers keep myriads of bits of information which are never used. Thus managers must be able to sort out the relevance of observations and discard the useless. This involves critical thinking. That is, the skill to look at information and decide whether it is accurate and relevant. How often have you been told in an advertisement, or popular article, that, say, a particular computer-based accounting system is absolutely essential for good management and control? If you believed all advertisements in their entirety, you would be buying a significant heap of inputs that would put you on the wrong place of the production frontier. Critical thinking should be a manager's constant companion.

Good observation requires:

- good listening;
- reading; and
- watching skills.

Sights and sounds constantly bombarding a manager must be noted, sorted, discarded or stored as the case may be. How good are you at being shown a table full of objects, and then remembering what you were shown? There are many books discussing how to improve your observation skills, so every manager should hone these attributes. But they must be good at not only visual observation, but also at picking up the relevant information when listening to people in one-to-one situations, or group situations, and also at picking up the relevant information when reading.

Listening skills do vary considerably, and can be learnt through a good text. While listening sounds like a passive affair, a good listener is actually practising what is called 'active listening'. This primarily involves feeding back to the speaker little summaries of what they have said to ensure you have picked up the central message. Encouraging, but brief, comments are also important to show the speaker that you are noting what is said and are supporting his or her efforts.

Similarly, reading skills vary between people. Research has shown a particular set of approaches is beneficial. This involves:

- skimming to start with so you can decide whether the material is relevant; and
- then starting back at the beginning picking up the main messages.

A good memory helps with these skills for you want to be able to store and retrieve the important and relevant information, or, at least, the essence of it and where the details can be found when needed. Again, there have been many books produced on how to develop memory skills.

No matter how good the memory, a manager will usually need some kind of recording system, even if just for tax purposes. Thus, the skills to:

- know what to record; and
- how to do this

are important. Essentially, records should be kept if they can be used for improving decisions provided the cost in time of obtaining and recording the records is not greater than the increased return from their use. That is, information has a cost and return so the manager should keep sufficient records such that the marginal cost equates with marginal return. Of course, the skill is in knowing this point!

In the record-keeping process, the use of a business computer is probably very relevant provided adequate integrated software is available. But remember some records will be kept for other than monetary reasons. With a strong interest in the environment, perhaps a manager keeps records of rural wildfowl activity purely for general enjoyment.

Finally, under the observation skill, a manager must always be on the lookout for new knowledge that might have some application on his business. This will involve:

- extensive reading;
- talking and listening;
- field day and 'in the field' visits to other businesses; and
- computer searching and watching.

Indeed, one assessment of the features of successful managers showed the one common thing between them was the time spent reading and searching out new knowledge, be it technical or financial related.

Negotiating

Third, the managers rated NEGOTIATING skills highly. Most managers will indeed be involved in many negotiations, from fixing prices through to setting contracts for new employees. Special skills and training are important to good negotiating. Some people spend their life as professional negotiators. Courses and texts on negotiation abound.

One of the central themes is that there is almost always a set of conditions that both, or more, parties will be content with. The skill is in using imagination to come up with a suitable package. A simple example is the manager who wants a certain price for the wheat-based products produced, but the buyer is offering somewhat less. Clearly both may need to alter their stance, but by how much depends on the competitors' positions. Nevertheless, perhaps there are other factors that will help. The manager might have some storage facilities which can be used if the buyer is struggling to keep all the grain he needs. So, a compromise might be a small drop in price together with the manager agreeing to hold the wheat in his storage for 2 months post 'sale'. A down payment might also be arranged.

In negotiations it is important to consider all factors that might impinge on the situation and work around variations on these until a meeting of the minds occurs. Of course, in some cases it will pay to walk away from a negotiation where the seller recognizes that someone else can offer a better deal.

Anticipation

Fourth, ANTICIPATION is a critical core skill, as recognized by the managers. Anticipation of what might occur is critical in ensuring all situations can be catered for. If opportunities are lost, or problems not corrected, the outcomes will be less than what is possible (in succession planning, delay may destroy opportunities. See Fig. 5.2). Anticipation is particularly important when planning because a manager must imagine what the outcomes might be from following certain decision paths. The whole of decision making revolves around anticipating what might happen as a result of taking alternative decisions. Without successful anticipation, decisions become a haphazard affair.

Successful anticipation depends totally on visualization. The managers, most of whom tend to be 'concrete' thinkers, must be able to mentally visualize what might happen this month, this year, next year and perhaps through

Fig. 5.2. Planning ahead is critical. Anticipation and succession planning must start very early.

to several years, as a result of taking possible decisions. The results of the visualization lead to calculating comparative budgets, even if mentally, giving the summary figures enabling decisions. An important component of visualization is imagination.

It is well known that some people have more active imaginations than others, but there are training packages that can help bolster imagination.

It is also interesting to note that most successful sportspeople are encouraged to use visualization, and there is clear proof that those who are good at it succeed. A goal shooter visualizes the ball arching away from her or his boot and entering the goal just out of the reach of any defenders. The high jumper has a mental picture of where her/his limbs will be as s/he clears the bar. The mental image is a bit like the TV replay, but in this case it is before the event.

Planning

The managers also rate PLANNING FOR THE SHORT AND LONG TERMS (fifth) as being important. This must involve making economic forecasts and comparisons that involve the visualizations of inputs and outcomes. However, after the decision on which alternative path to take, the plan then becomes the blueprint for action. What has to be ordered? And when? As without the inputs on hand at the right time, and the means of implementing the tasks, success will be less than optimal.

Learning from experience

Sixth, LEARNING FROM EXPERIENCE was a skill the managers thought very important. There is unlikely to be any industry where the ability to improve management skills from the lessons of experience is not important. Everyone has to start somewhere. Textbooks and courses are not a substitute and cannot provide an appreciation of all the factors involved in the decision making, implementation and control cycle.

Despite the importance of experience, little research has been conducted on the factors important in gaining experience. The intrinsic skill of self-observation is probably important. To learn you need to analyse what was carried out, what went wrong or succeeded, as the case might be, and how improvements might be made for future situations that mirror the experience. However, it must be recognized that seldom do exactly the same situations repeat themselves in most rural situations, nor in most other business arenas, so learning from experience will involve the ability to intelligently note the differences of the current situation and modify the experience of the past to suit. Recognized experts are people that have learnt from their experience. There is also some luck in experience in that one manager may have had all the right experiences in the past in contrast to another who may never have experienced, say, a season that is twice as wet as normal.

People skills

PEOPLE SKILLS is undoubtedly an important attribute, and this was clearly recognized by the managers. As the seventh core skill the managers believed successful judging of personality, resolving conflicts and good relationships of the business were all important aspects of 'people skills'. People skills are also no doubt related to negotiation skills. In that a business operates in an environment of people, being able to communicate successfully is all important, as is a tolerance of different views and ways, provided the job gets done. Being able to maintain friendly relationships with neighbours, consultants, bankers and so on is crucial to success. Again, there are many books and training courses on people skills that can help develop these skills.

Implementation

Eighth, IMPLEMENTATION OF PLANS is a critical skill as recognized by the managers.

Optimal plans are no use if they cannot be successfully implemented. These two links in the chain must be as equally strong. As you would expect, the managers commented that 'acting in time' was critical, as was 'not panicking'.

Research has shown that managers experience considerable stress relative to other occupations due to the lack of control they have over many of the conditions. Given a series of bad outcomes it is important to maintain logical decision making despite the pressures. In this respect, a manager's personality is an especially important factor. 'Acting in time' requires little comment – particularly where the business is in a biological world as processes cannot be turned on and off at will. The same applies to markets. It is clear that it is the early bird that catches the worm on both accounts.

To this might be added the importance of 'doing the right thing here and now'. Most production processes are dynamic and fluid, requiring constantly changing action to suit the circumstances. Equally, decisions should not be made ahead of time and rigidly maintained for they may no longer be appropriate if, for example, the market saturation level is different from the expected. A manager must have a range of strategies in his or her 'kit bag', each one of which is used according to the particular condition set. Waiting to the last minute to see the conditions is equally as important as not delaying action. It is a fine balance that makes successful production and marketing such a fascinating study and challenge.

Technology

The managers also recognized that UNDERSTANDING THE TECHNOLOGY was critical in that they rated highly 'understanding interrelationships'. This ninth core skill is patently obvious for without it decision making becomes nothing but a lottery. To this must be added TECHNICAL SKILLS, for without these, implementation becomes a rather hit-and-miss affair. In that many business and production systems involve many disciplines in all their complexity, the art of simplification becomes very important. It is essential to be able to isolate the important and essential components of the technology and its interrelationships to reduce the decision problem to manageable proportions. It will be recalled a feature of experts was their ability to simplify, in a meaningful way, the complexity of a problem.

Solutions

What the managers did not mention at all was the skill of creating realistic solutions to problems. This might be called the SOLUTION GENERATION skill (tenth). It is all very well to recognize problems and opportunities, but to create the optimal solution, particularly where it is new and unique, is rather more difficult. This skill requires an intimate understanding of the technology, markets and production conditions as well as creativity and imagination. Many managers lack the latter two attributes, largely because they have not been encouraged to be creative. There are training programmes which promote thinking of new solutions and processes.

Analysis

Finally, the eleventh core skill is ANALYTICAL KNOWLEDGE. The manager can be very creative, but suboptimal outcomes will eventuate if the manager does not understand how to analyse the alternatives both in a tactical and a strategic sense. An outline of production economics principles was given earlier. These must be used if the right choices are to be made. A knowledge of what constitutes a fixed cost relative to variable costs is essential in deciding action, as is the concept of 'opportunity cost' in which it is necessary to use the value of a resource in its next best use when looking at the return from an action which takes away resources from other uses. This assumes that it is possible to measure in some way the 'business's' objectives because this measurement must be the yardstick used. This is difficult, so the practical approach of using monetary costs and returns is often used with subsequent subjective adjustments for other objectives. Thus, the alternatives might be ranked on their financial returns, and then possibly re-ranked once their contribution to leisure time, enjoyment, complexity and so on is subjectively taken into account.

With the development of management software, computer packages are becoming available for helping ensure that the correct analytical approach is used. Similarly, packages that will determine practical optimal courses of action are slowly appearing. There are also many packages that help keep records which can turn data into information, the distinction being that information can be used directly for decision making. For example, a financial package that records all transactions, if properly designed, can produce enterprise profits which might then be projected into the future after allowing the manager to update the past information to reflect the outlook.

Case study managers' views

The case study managers struggled a little when asked what they believe were the most important skills. They probably just do not think in these terms, preferring to just 'get on with the job'. Making decisions does not require them to make such lists.

Margrave certainly knew which decisions he found the hardest, even though in general he noted 'I have never found making decisions particularly hard ... they just got made'. For the rare difficult decision he was not comfortable with, and spent some considerable time thinking about, once made he simply moved on to the next problem. In general, he believed staffing decisions were the most difficult, perhaps because humans cannot be quantified as easily as other factors. 'Appointments, sackings and correcting staff' were all difficult. One suspects that most managers feel the same as on a business you work intimately with the staff.

Generally, Margrave believed important skills and attributes were:

- Knowing when to act.
- Having a strong desire to be a good manager, to strive to improve.
- Being able to enjoy working with people.
- Good communication skills.
- Having an enquiring mind.
- An ability to ask someone else for another opinion, or for information and ideas.
- An ability to constantly watch for new information ('awake enough to tap into the information available').
- Good anticipation skills, like constantly doing cash budgets to anticipate deficits and surpluses and making provision for each as needed.
- Credibility with labour. The manager must be able to perform all tasks successfully, or at least fully understand what they are all about, and consequently have the staff respect and admire the manager.

Having provided this list Margrave also commented: 'Good profits are almost a chance factor ... if no floods occurred, nor a market collapse occurred, then profits were good as the business was normally producing at high levels. Sudden upswings and downturns in the market also had an impact as in such cases forecasting was impossible. So often the political decisions on the other side of the world impacted.'

When talking about staffing issues, Margrave wanted to make the point that it was very important to work with staff and give them responsibility as well as opportunities to take time off to attend short courses. Each staff member had his own area of responsibility and was generally left to get on with that area. The non-threatening approach Margrave took when helping and watching over outcomes seemed to be appreciated by the staff. Staff turnover was minimal, reflecting the success of Margrave's approach and management. Margrave believed 'most people want to do the right thing', so providing training and praise for good outcomes gave rise to a contented and successful set of workers who more than repaid the time and money spent on them.

Overall, when asked which skills were important to his success, his reply was 'I'm successful because I enjoy the business type and the manager's job. I was not much use at school, but the moment I left school I couldn't get enough information about business and management. The journey is as much fun as getting there', stressing that enjoyment is high on Margrave's agenda. He also believes: 'I'm quite creative with a good imagination – a critical skill. You must imagine and write down your thoughts.'

And certainly Hank writes a lot. He makes lists of the jobs that must be accomplished in priority order, and constantly reviews both what to keep on the list, and in what order. Hank is also a firm believer that experience is the key to success, or rather, as he puts it 'technical knowledge + natural ability + experience gets the right jobs done'. He adds 'to interpret experience you need experience to understand why'. Hank's list, then, includes:

- The ability to mull over experiences and learn the lessons on offer.
- An ability to read a lot and absorb technical information for general use.
- Good labour management skills.
- Good skills at using spreadsheets and doing the sums on alternatives.
- An ability to judge the outcomes from changing the business and production systems.
- The confidence in your conclusions to act decisively.

Hank believes he learnt many lessons on managing labour from early mistakes. Initially Hank tried to direct his helpers with detailed instructions, but soon learnt many resented such specific control. Hank learnt you had to 'give employees scope'. It is interesting that Margrave quickly came to the same conclusion, which ended up with him giving each person an area of responsibility.

With the volatility of prices and conditions, Hank believes it is important to constantly update cash flows and budgets so you are aware of the changing bottom line and bank overdraft. Quick decisions can then be made about problem situations. It is no use shutting the gate after the horse has bolted. This belief is reflected in his list of skills, and certainly with developing a new business involving vast investments good control of the cash flow situation is imperative.

During his time in a rural bank Hank learnt to quickly assess a proposal and decide if, and how much, the bank would lend. This experience has stood him in good stead. Hank notes the toughest decision he has faced in recent times is which milk processing company to contract to for his new business. Corporations are not always totally forthcoming with their information, for, no doubt, it helps to keep some of the details somewhat hidden. In the end, after doing many long-term budgets and world market analyses, Hank decided to go with the cooperative with which he was familiar. Perhaps this reflects his risk-averse nature. He concluded 'that was a real hard decision'.

Introducing Two Consultants, and Their Views on Skills

Bruce and his dairy business interest

Bruce really wanted to be a manager, but his access to the necessary resources precluded this option when he was young, and now that purchasing a business would be an option, Bruce believes he would find some of the physical jobs required rather boring, but he does enjoy the decision-making side of production. This is one of the reasons he is pleased he followed the path into consultancy. But to satisfy his almost inherent desire to be involved in practical business operations he has a small part-time business.

Bruce was brought up in a city, though his family had country friends where he spent as much time as possible. While his father was a city business person with little interest in rural activities, Bruce believes some of the urge to be a manager comes from his business ancestors in Scotland. This primary production fascination led Bruce to take agriculture at university where again most of his friends were from business operations, so holidays saw him visiting one or another. His best friend was from a dairy enterprise so perhaps these experiences sparked the keenness to be involved in this industry. Bruce even emulated managers by wearing typical manager attire.

Bruce believes he is risk averse as otherwise he might just have scraped together what funds he could find and started business acquisition. Instead the university path was followed, and eventually postgraduate study. A strong involvement in the Boy Scout movement had Bruce taking many leadership courses and ending up as one of the leaders. This training, and perhaps his father's interests, led Bruce into management and economics courses to satisfy this fascination with organizing, management and decision making. As Bruce comments: 'I was always intrigued with how to put resources together and make money.' This interest also stemmed

Fig. 5.3. This is where Prof's journey into food and fibre business started.

from his romantic visions of being a manager, but the family and friends reckoned a city boy could never join the rural business community. They were wrong.

Bruce left university and went straight into consulting positions, and on reflection wonders how he managed so well given his lack of experience particularly after making an initial mistake which knocked his confidence. However, he did overcome problems, and successfully related to managers, becoming, in the end, a strongly respected consultant dealing with both individual managers as well as groups over technical issues in the dairy enterprise business.

When giving his views on the skills successful managers require, Bruce comments that business skills is not something which you pick up overnight. To Bruce, as a newcomer who has taken a long time to absorb the unspoken nuances, this is an important observation. Bruce noted, in talking about business family dynasties, 'succession is not just the business, but all the skills and heritage that get absorbed almost by osmosis'.

'There is heaps of stress in rural business' is an observation that leads Bruce into noting an important skill is handling risk. 'A crop man invests thousands into cultivation, seed, and fertilizer in the hope that it will rain.' Who else would have such confidence? This is one reason why good managers 'must not be scared of debt ... for without borrowing little progress can be made'.

A manager must also be an assiduous observer noting all changes, and quickly responding appropriately. 'You can't just stay static.' And in this process success requires the manager to have excellent decision-making skills with an ability to judge the information being received, analyse it and act accordingly. In a nutshell, the manager must be good at responding to change.

Bruce also believes managers must be totally open-minded and assess each situation on its merits for improving their skill is dependent on learning from mistakes. Thus, an important skill is being able to say 'Hey, I've made a mistake, how can I learn from this?'

This 'willingness to learn creates a good manager'. And 'if the manager exposes himself to stress some learn and improve, others sink'. This is all part of 'being competitive and resilient' and not wanting 'other managers to beat me'. Bruce comments that the 'personal factor' is vital to good management. By this he means the manager's honesty, integrity and work ethic: 'Can he meet a challenge, and go the extra mile?'

Thus, a manager's competitive nature is an important attribute in driving him to improve, 'do they want to get respect, or just plod on?' Bruce comments that a manager's entrepreneurial skill is part and parcel of success. If you expose them to possibilities, 'some pick them up, others don't. They need this innate ability to start with. It is very difficult to help someone without this desire.'

A manager must have strong prioritization skills. Bruce talks about a quadrant of jobs with each being given a grading based on how important they are, and the urgency with which they must be attended to. Thus, you have non-important jobs with little urgency through to very important jobs that require to be completed with urgency. It is only this latter grouping that should be attended to. Hank the dairy manager clearly has this prioritization skill, using his constantly updated diary to sort out the jobs in this 'to do' quadrant.

No doubt, given more time, Bruce could have added to his list of skills, but the list discussed is what stands out in his mind as being associated with high managerial skill.

'Prof' and his comments about the skills required

The second consultant we will call 'Prof', largely because he has been involved in a wide range of activities within business and systems analysis including working as an academic and researcher. This wide range of experience means he has come to well-based conclusions on managers and their modus operandi through many different eyes. Prof started his professional career as a 'business adviser', as they were called then, but today he would have been called a consultant. This career then moved into research, and, finally, tertiary teaching and research student supervision. Through the years Prof has also been involved with international consulting, some of which was helping tertiary establishments develop business management teaching. Prof has also supervised a research business that attracted hundreds of managers interested in the groundbreaking systems developed. In a nutshell, Prof is eminently qualified to comment on what makes a successful manager.

You might imagine Prof was stimulated to become involved in primary production systems through his early experiences and contacts, but nothing could be further from the truth. Prof was a city boy who had no contacts at all in rural areas, and his relatives and recent ancestors have all been urban people. One wonders where the strong desire to be involved in primary production came from. As Prof notes: 'I just had this strong interest even though I went to a high school that was only interested in the traditional professions ... engineers, doctors, lawyers, accountants.' Prof's father was an engineer, who was very interested in his son's activities, development and career, and whose only stipulation was that Prof should go to university (Prof's first university was primary production focused. See Fig. 5.3) rather than work in business operations. In fact, Prof's first contact with consultancy was spending 3 days with a business adviser to gain insight into what was involved. He organized this contact himself while still at school, so there was a strong motivation stemming from some inner source. Prof's first practical experience of rural business was spending the summer holidays immediately before moving to university in such a business. An interest in the technology involved has never left him, and he still takes a strong interest in the development of the industry.

The other great love Prof immersed himself in was mountaineering. While still at school he joined a tramping club and enjoyed their activities, noting that 'I had this great attraction to the wide outdoors', and perhaps this was part of the desire to be in rural business operations. The mountaineering love led Prof to take time off work and education to visit a wide range of mountain ranges including high peaks in South America, the Himalayas and Antarctica. The challenges and demands of high mountaineering clearly developed Prof's determination, survival abilities and people skills, which, in turn, helped his professional career. Prof comments: 'One of the reasons for doing postgraduate study was the longer holidays which enabled following my passion for the mountains.'

Today Prof leads a very busy life teaching, writing, addressing different groups including managers, and international consulting. However, this schedule now precludes spending much time in the mountains. The price of success is sometimes hard to take, but through all this, Prof has clear views on the skills a manager should have if he is to be similarly successful.

Top of the list is technological knowledge and expertise. 'This goes without saying', he comments, for it is the foundation from which success occurs provided successful application is achievable. The main other key is being able to put the technology into workable systems. The business is an integrated whole of many connected components, so it is important to create from the jigsaw parts something that achieves the objectives. A strong curiosity is one of the skills that enables the successful systems to be put together, a strong curiosity to search out new ideas and ways of integrating them into a whole. Prof believes a manager faces one of the most complex decision set-ups existing in modern society, and creating a successful system is challenging, but something that managers in general do well. In this respect Prof holds managers in extremely high regard.

An important skill, according to Prof, is the ability to observe what other managers are doing and consequently weeding out the better systems which can then be applied 'at home'. Thus, networks are important, and benchmarks which the manager has in mind so that he can assess the possibilities and weigh them up against his current system. An ability to create a good team about them is an important skill enabling an assessment of all the ideas and their integration into a good total system. While the manager might not have all the skills themselves, Prof notes that 'the ability to know your strengths and weaknesses leads to selecting a team that together forms a complimentary package'.

Agreeing with the other managers and consultant, Prof stresses that in any business the successful manager must be passionate about the type of business and production systems. If he is not, he just won't succeed, for business is nothing like an 8-to-5 job after which you move on to other parts of your life. Prof comments that 'a good manager will be constantly mulling over all the options, strategies and tactics no matter where he is in an effort to come up with the best decisions and actions. This requires dedication, total commitment, and this passion for being involved in the chosen production system and business.'

One of the core reasons for the need for constant thinking is the risk and uncertainty that surrounds most production systems. 'You can't make a decision then walk away from it for something is bound to change requiring a re-evaluation.' Thus another important skill is the ability to be flexible. As Prof puts it, 'you need to get yourself into a space where opportunities arise as conditions change'. This is partly a frame of mind as well as having a range of options at your fingertips. However, Prof comments, 'this doesn't mean you don't plan. In fact planning is everything as going through the planning process puts you in a position to take the opportunities.' You need to 'think through the strategies and tactical responses for a manager works in a complex uncertain environment'. However, the best approach varies with the type of business. Some rural production systems are akin to a town factory, examples are pig and poultry production as well as feed lots. In these situations, production is largely isolated from the weather, so an important skill is being able to plan systems over long periods of time.

Good management requires the manager to be self-aware and critical. Prof notes that a manager must examine all actions with a strong yardstick using sensible benchmarks. In other words, complacency is an enemy, and 'the arrogant manager won't make much progress'. Thus, an open mind is an important attribute preventing preconceptions from blocking a proper examination of alternatives and opportunities.

As noted, Prof is a strong believer that a manager should be fully aware of his personality so that he understands his strengths and weaknesses. Prof recounts the use of standard tests to enable members of a management team, sometimes a husband and wife, to examine how they fit together and, therefore, where they may need to make special efforts to support and control behaviour. Prof also talks about the use of the Margerison–McCann Team Management Wheel (https://www.mindtools.com, accessed 27 August 2024; www.tms.com.au/tms07.html), in which people are classified into eight types. These are the reporter–adviser, creator–innovator, explorer–promoter, assessor–developer, thruster–organizer, concluder–producer, controller–inspector and upholder–maintainer. The right mix can enhance a team, but even if a good team does not fall out given the people involved, at least an understanding of each type helps people adjust.

In a nutshell, Prof believes the following skills and attributes are all important to success in business management:

- An excellent technological knowledge and ability to implement the chosen systems.
- An ability to develop integrated whole-business systems.
- Having a strong curiosity.
- Successful at observing what other managers are doing, and judging what is observed using benchmarks.
- An ability to create a team that complements each other.
- Knowing and understanding your own personality, strengths and weaknesses.
- Having a strong passion for the chosen production and type of business.
- An ability to constantly review outcomes, options and alternatives.
- An ability to be flexible in contrast to preferring to work with fixed systems.
- Self-awareness and an open mind.

Can you add to this list? Prof also talked about the need for labour management skills now that many dairy businesses in particular are employing as many as 12 people. Some managers find this number difficult to handle. One manager Prof talks about found his personality just did not fit in with large numbers of employees, so he cut back his operation to five people as he could then work his favoured personal mentoring system with each labour unit. Each was much happier and worked well together.

Concluding Comments

There are major differences between managers in their managerial skill. One wonders why these differences occur, and what can be done to improve the skills of all managers, good and bad. It is probably true that most skills are learnt from experience, often before becoming a manager, especially where the manager has lived in rural areas on a business from a young age. Thus, most managers grow into the management role and will have a full range of skills, whether good or not so good, through this process. However, this chapter has highlighted the core skills that all managers will have even if they cannot list them out themselves. This list enables a systematic consideration of the components of a manager's total skill, and which sub-skills might need extra attention. Similarly, the list is a blueprint for the skills that need including in any formal courses on improving management.

The ranking of the skills was obtained from a large group of sophisticated and well-educated managers involved in businesses valued at around US$5,000,000 on average. However, it would be surprising if the skill list would be much different in less-developed business situations. One area that might be more important is 'people skills' due to the wider involvement of family and community as shown by studies quantifying decision processes in less-developed countries.

It is also important to note that the type of business did not seem to impact on the core skill list, though this would be expected as details of, for example, the particular technology knowledge required was not investigated, nor the specific technical skills.

Finally, it is worth noting that one of the highly ranked specific skills was 'having an adventurous spirit'. In general, many rural operations are certainly other than a controlled factory situation where processes and procedures are standard fare. The challenges and complexities are such that people with an adventurous spirit are more likely to succeed.

5A. Survey on Managerial Factors

Listed below is a copy of a survey used to obtain manager's views which came from a base for developing more surveys in other domains. It started with the instructions ...

Please complete and return this questionnaire using the enclosed postage-paid envelope. All information provided will be kept in strictest confidence to the researchers involved. If you are not the operator or manager of the business please pass the questionnaire on to this person.

A. General

1. **Business type**. Please tick ONE box representing the MAJOR enterprise type on the business you operate. Herewith list possibilities with tick boxes ...
2. **Labour**. Including the manager, please give the number of equivalent full-time adult people it takes to run the business (use fractions if necessary, e.g. three-fourths). ☐
3. **Size of business**. What has been the average gross annual turnover during the last 3 years? ($) ☐

B. Importance of Managerial Attributes

For your situation, please rate the importance of each of the managerial attributes listed below. Use a score range of **7** (VERY important) to **1** (NOT AT ALL important) with four representing MODERATELY important and the other numbers for in-between degrees of importance.

1. Ability to identify the key factors in a problem and discard the irrelevant. ☐
2. Quickly analysing and sorting out situations that have never been faced before. ☐
3. Having a clear understanding of the family's objectives, values and goals, thus making assessing the value of alternative actions easy. ☐
4. Being able to predict local weather better than the official forecaster. ☐

5. Understanding the local political scene as it might impact on rules affecting what can be done.

6. Developing and maintaining a support network of colleagues and professionals.

7. Being able to efficiently organize and carry out quite complex operations (e.g. get a new packing shed operational on time).

8. Developing appropriate and detailed plans for both short- and longer-term horizons.

9. Making requirements clearly understood (effective communication).

10. Understanding the basis on which to choose between alternatives (e.g. knowing how to cost unpriced labour, knowing how to do gross margins, understanding diversification principles, etc.).

11. Being up to date with the current condition of the business in its totality (bank balances, resource conditions, production levels, asset lists, machinery repair, etc.).

12. Picturing (understanding) the consequences of a decision over the many (or few) months/years it might impact over (e.g. planting an area in forestry, installing new production set-ups, etc.).

13. Skill at keeping, interpreting and using recorded data about the business and associated factors (e.g. market trends).

14. The ability to predict product prices into the foreseeable future, or at least understanding the factors that determine the prices, and understand market requirements.

15. Assessing job priorities.

16. OTHER – If you think an important managerial component has been left off the list, please write it below and give it a score.

(i) ________________

(ii) ____________________

C. Importance of Entrepreneurial Skills

For your situation, please rate the importance of each of the entrepreneurial skills listed below. Use a score range of **7** (VERY important) to **1** (NOT AT ALL important) with four representing MODERATELY important and the other numbers for in-between degrees of importance.

1. Being able to seek out, identify, and clarify new opportunities (production, products, marketing, etc.).

2. An ability and determination to look/ask/seek out information thought to be necessary for making decisions.

3. Ability in learning new skills.

4. An intuition that gives early-warning signs when something is right, or, in contrast, when something positive needs exploiting. ☐
5. Skills in finding the very best market (price, quantity, etc.) for all output. ☐
6. Understanding deadlines and being able to 'act in time' (e.g. spray before insect damage, fertilizer applied in good time). ☐
7. The skill to negotiate the best possible deal (price, arrangement, etc.). ☐
8. A belief in being able to control a lot of what happens around the business; in contrast to a belief that not much is really controllable due to the weather, markets, government action, etc. ☐
9. When faced with opportunities, ensuring ALL alternatives are sought out, considered and evaluated. ☐
10. The skill and intuition to forecast well into the future likely opportunities in products and production systems. ☐
11. An ability to look ahead and anticipate likely problems, needs and opportunities. ☐
12. Understanding sources of risk and what can be done to reduce its impact. ☐
13. OTHER – If you think an important entrepreneurial component has been left off the list, please write it below and give it a score. ☐

(i) ________________ ☐

(ii) ________________ ☐

D. Managerial Style

For each of the following statements indicate how true it is with respect to your management style. Each question has five boxes beside it – tick only the ONE that best records the degree of truth in the statement.

1. You tend to mull over decisions before acting.

TRUE ❑ ❑ ❑ ❑ ❑ NOT TRUE

2. You find it easy to ring up strangers to find out technical information.

TRUE ❑ ❑ ❑ ❑ ❑ NOT TRUE

3. For most things you seek the views of many people before making changes to your operations.

TRUE ❑ ❑ ❑ ❑ ❑ NOT TRUE

4. You usually find discussing everything with members of your family and/or colleagues very helpful.

TRUE ❑ ❑ ❑ ❑ ❑ NOT TRUE

5. Where there are too many jobs for the time available you sometimes become quite anxious.

TRUE ❑ ❑ ❑ ❑ ❑ NOT TRUE

6. You tend to tolerate mistakes and accidents that occur with employees and/or contractors.

TRUE ❑ ❑ ❑ ❑ ❑ NOT TRUE

7. You share your successes and failures with colleagues.

TRUE ❑ ❑ ❑ ❑ ❑ NOT TRUE

8. Keeping records on just about everything is very important.

TRUE ❑ ❑ ❑ ❑ ❑ NOT TRUE

9. You admire business/production colleagues who are financially logical and don't let emotions colour their decisions.

TRUE ❑ ❑ ❑ ❑ ❑ NOT TRUE

10. You sometimes don't sleep at night worrying about decisions made.

TRUE ❑ ❑ ❑ ❑ ❑ NOT TRUE

11. You find investigating new business/production methods exhilarating and challenging.

TRUE ❑ ❑ ❑ ❑ ❑ NOT TRUE

12. You tend to write down options and calculate monetary consequences before deciding.

TRUE ❑ ❑ ❑ ❑ ❑ NOT TRUE

13. You tend to worry about what others think of your methods.

TRUE ❑ ❑ ❑ ❑ ❑ NOT TRUE

14. You are happy to make do with what materials you have to hand.

TRUE ❑ ❑ ❑ ❑ ❑ NOT TRUE

15. You find talking to others about business/production ideas stimulates and excites you as well as increasing your enthusiasm for new ideas.

TRUE ❑ ❑ ❑ ❑ ❑ NOT TRUE

16. Having to make changes to well-established management systems and rules is a real pain.

TRUE ❑ ❑ ❑ ❑ ❑ NOT TRUE

17. You normally don't rest until the job is fully completed.

TRUE ❑ ❑ ❑ ❑ ❑ NOT TRUE

18. You normally enjoy being involved in manager/production organizations.

TRUE ❑ ❑ ❑ ❑ ❑ NOT TRUE

19. You sometimes believe you are too much of a stickler for checking and double-checking that everything has been carried out satisfactorily.

TRUE ❑ ❑ ❑ ❑ ❑ NOT TRUE

20. When the pressure is on you sometimes become cross and short with others.

TRUE ❑ ❑ ❑ ❑ ❑ NOT TRUE

21. You generally choose conclusions from experience rather than from hunches when they are in conflict.

TRUE ❑ ❑ ❑ ❑ ❑ NOT TRUE

22. You are inclined to let employees or contractors do it their way.

TRUE ❑ ❑ ❑ ❑ ❑ NOT TRUE

23. You not only speak your mind and ask questions at manager/production meetings, but also enjoy the involvement.

TRUE ❑ ❑ ❑ ❑ ❑ NOT TRUE

24. It is very important to stick to management principles no matter what the pressure to do otherwise.

TRUE ❑ ❑ ❑ ❑ ❑ NOT TRUE

25. You are much happier if everything is planned well ahead of time.

TRUE ❑ ❑ ❑ ❑ ❑ NOT TRUE

E. Goals and AIMS

For each of the following statements indicate how true it is with respect to your goals and aims. Each question has five boxes beside it – tick only the ONE that best records your degree of belief in the statement.

1. It is very important to pass on the business to family members.

TRUE ❑ ❑ ❑ ❑ ❑ NOT TRUE

2. It is important to earn the respect of managers/producers in the local community.

TRUE ❑ ❑ ❑ ❑ ❑ NOT TRUE

3. Making a comfortable living is important.

TRUE ❑ ❑ ❑ ❑ ❑ NOT TRUE

4. It is very necessary to keep debt as low as possible.

TRUE ❑ ❑ ❑ ❑ ❑ NOT TRUE

5. It is essential to plan for reasonable holidays and plenty of leisure time.

TRUE ❑ ❑ ❑ ❑ ❑ NOT TRUE

6. Attending 'field' days and manager/producer meetings is vital.

TRUE ❑ ❑ ❑ ❑ ❑ NOT TRUE

7. It is very important to reduce risk using techniques like diversification, managing/operating conservatively, keeping cash reserves

TRUE ❑ ❑ ❑ ❑ ❑ NOT TRUE

8. Developing facilities and systems that give good working conditions is crucial.

TRUE ❑ ❑ ❑ ❑ ❑ NOT TRUE

9. It is very important to ensure employees enjoy their jobs.

TRUE ❑ ❑ ❑ ❑ ❑ NOT TRUE

10. Doing jobs that I enjoy is a very important part of the operation.

TRUE ❑ ❑ ❑ ❑ ❑ NOT TRUE

11. Minimizing pollution is very important.

TRUE ❑ ❑ ❑ ❑ ❑ NOT TRUE

12. I enjoy experimenting with new products and production systems.

TRUE ❑ ❑ ❑ ❑ ❑ NOT TRUE

13. Proper retirement planning is a major consideration.

TRUE ❑ ❑ ❑ ❑ ❑ NOT TRUE

14. You must always be striving to increase the total value of assets.

TRUE ❑ ❑ ❑ ❑ ❑ NOT TRUE

15. Constantly expanding the size of the business is absolutely necessary.

TRUE ❑ ❑ ❑ ❑ ❑ NOT TRUE

16. Aiming for maximum sustainable net cash returns is very important.

TRUE ❑ ❑ ❑ ❑ ❑ NOT TRUE

17. Maintaining a presence in local community activities is important.

TRUE ❑ ❑ ❑ ❑ ❑ NOT TRUE

18. It is very important to improve the condition of the business (worker stations, facilities, etc.).

TRUE ❑ ❑ ❑ ❑ ❑ NOT TRUE

19. Giving assets to the children so they can pay for education and/or set up businesses is very important.

TRUE ❑ ❑ ❑ ❑ ❑ NOT TRUE

F. Computer use

If a computer/s is used for business in your enterprise, give the average HOURS PER MONTH that it is used for the following (otherwise go to the next question).

Recording financial transaction information	☐
Doing forecast budgets/cash flows	☐
Keeping machinery records	☐
Keeping production records	☐
Word processing	☐
Searching the Web for information	☐
Sending e-mails	☐
Entertainment/education	☐

Internet banking

Internet purchasing

OTHER ______________

G. Importance of Personal Attributes

For your situation, please rate the importance of each of the personal attributes listed below. Use a score range of **7** (VERY important) to **1** (NOT AT ALL important) with four representing MODERATELY important and the other numbers for in-between degrees of importance.

1. Early observation of important indicators around the business (e.g. machine breakdowns, market failures, labour worries, and so on).

2. Keeping a cool head and putting aside any tendency to panic when faced with stressful situations.

3. Having the confidence to draw conclusions and act quickly and decisively.

4. An excellent knowledge of facts, figures, procedures and methods, with respect to the technology of the business, all production processes and facility issues – machines, buildings, etc.

5. Being prepared to give it a go and take risks in changing production systems and/or starting new ventures.

6. High motivation in constantly seeking better ways and implementing them; in contrast to being happy with current systems.

7. Accepting the good and the bad, and not letting it affect management and decision making.

8. Ability to learn from experience, mistakes and failures.

9. The determination to keep working all hours until the high-priority jobs are completed.

10. Developing a 'good moral character' involving openness, integrity, reliability, trustworthiness, etc.

11. Developing a strong personality so that others' 'sit up, notice, respect and act' on what is said.

12. Understanding the interrelationships between all the components of the business (e.g. markets, inputs, machine and labour productivity, i.e. what affects what?).

13. Obtaining employees' and/or contractors' cooperation and understanding leading to harmonious and productive relationships.

14. Tertiary education in areas related to the business's technologies (e.g. type of business and its production and marketing, etc.).

15. Having above-average intelligence and school grades. ☐

16. Successfully judging personality and selecting suitable employees. ☐

17. Successfully resolving conflicts on, and off, the business (e.g. dispute between employees). ☐

18. Maintaining good relationships with outside people – bankers, accountants, suppliers. ☐

19. OTHER – If you think an important personal attribute has been left off the list, please write it below and give it a score. ☐

(i) ________________ ☐

(ii) ________________ ☐

H. Personal Features

1. Which age group do you fall into? (Tick ONE box)

Less than 25 years ☐ 26–35 years ☐ 36–45 years ☐
46–55 years ☐ 56–65 years ☐ Greater than 65 years ☐

2. What was the level at which you stopped your formal education? (Tick ONE box)

Primary school ☐ Secondary school – up to 3 years ☐
Secondary school – four or more years ☐ Tertiary education – up to 2 years ☐
Tertiary education – three or more years ☐

3. Please indicate your gender by putting **F**(emale) or **M**(ale) in the box. ☐

4. Please rate yourself in general intelligence. (Tick ONE box. If you are uncomfortable answering this question, leave blank)

Highly intelligent ☐ Reasonably intelligent ☐ Average ☐
A bit below average ☐ Other ☐

5. If all managers were rated on a **10** (excellent) to **1** (poor) scale for managerial ability, what level of skill rating would you give yourself? ☐

I. Managerial Training

1. To what degree would you use a managerial skill training programme, if available in your area? (Tick ONE box)

Not at all ☐ Occasionally ☐ Extensively ☐

2. Assuming training was available, please rank the following method of delivery in order of preference (one for most preferred, **2** for the second preferred, and so on).

Computerized self-training ❑ Book-based self-training ❑
Tutored system based locally ❑
OTHER (please specify) ____________________

3. On what other topics/skills would you like training?

(i) ____
(ii) ____
(iii) ____

6 Biases and Stress

Abstract
Any decision that varies from the best decision for a particular situation is referred to as a biased decision if the error occurs consistently. A manager must look out for such biases and make corrections. This chapter reviews the common biases, providing help for recognizing possibilities. The discussion categorizes bias types into observational, forecasting, decision process, decision implementation and also 'general'. For each the typical biases are defined. Examples include the 'halo effect', the 'anchoring effect', 'risk estimate failures', 'recency effects', 'chunking errors', decision 'timeliness' and 'aversion to change', and many more. Examples of each type are provided. Furthermore, making decision errors can be a source of stress to include in the list of the many more reasons why manager stress can occur. The last half of the chapter discusses the sources and covers methods of dampening stress where it is impacting on the operation of a business and manager well-being. The case study managers' thoughts on biases and stresses conclude the chapter.

Introduction

The majority of managers do not get it right on all occasions. Mistakes are made in one or more of the attributes or skills that are involved in a good decision. These skills were listed in the last chapter (Chapter 5). If, however, a manager is consistently wrong in some of the skills, then his decision making can be said to be *biased*. For improvement, the bias must be recognized, and the error of the manager's ways corrected, so that in future mistakes are only random in contrast to consistent. This chapter contains a description of the more common biases, providing comments on their recognition and correction. Of course, in many respects the full list of potential biases has already been given in that a consistent aberration in any of the skills required is a bias. However, it is still useful to highlight the common biases that researchers have particularly noted.

Then there is the problem of stress. Business, production, marketing and labour issues involve considerable risk and uncertainty and, consequently, give rise to considerable stress. There is nothing more disheartening than planning for developments, committing a lot of money, only to find that the project fails due to, say, a particularly poor year climate-wise or a natural disaster. The feeling of a lack of control can be overpowering. It is not uncommon to see reports of manager suicides under extreme situations. Stress, then, is common, but what matters is how it is handled and whether it impacts on the effectiveness of a manager's skills. A section of the chapter covers types of stress and what might be done to help remove the impacts of stress.

It is not yet exactly clear what creates a bias in a manager, and why a manager has excessive stress, but clearly the manager's personality, intelligence, family background and experience must all impinge on the creation of biases, and stress. The education system must also give rise to some of the biases. No one wants to acquire biases, nor overbearing stress, and so in most cases their existence is due to oversights of various kinds. It must be remembered that everyone makes mistakes from time to time, particularly due to chance elements (it is always easy to see what should have happened in hindsight, but no one has a perfect prediction record), so care must be taken in deciding whether a bias exists, or whether, on average, the correct approach is being used. Similarly, some stress is normal, and indeed can lead to careful and correct decision making, but if the stress is too much, decision efficiency declines, and certainly enjoyment also declines similarly.

Biases usually relate to decision making in contrast to the physical activities around a business. Clearly, however, physical prowess, or

DOI: 10.1079/9781800627246.0006

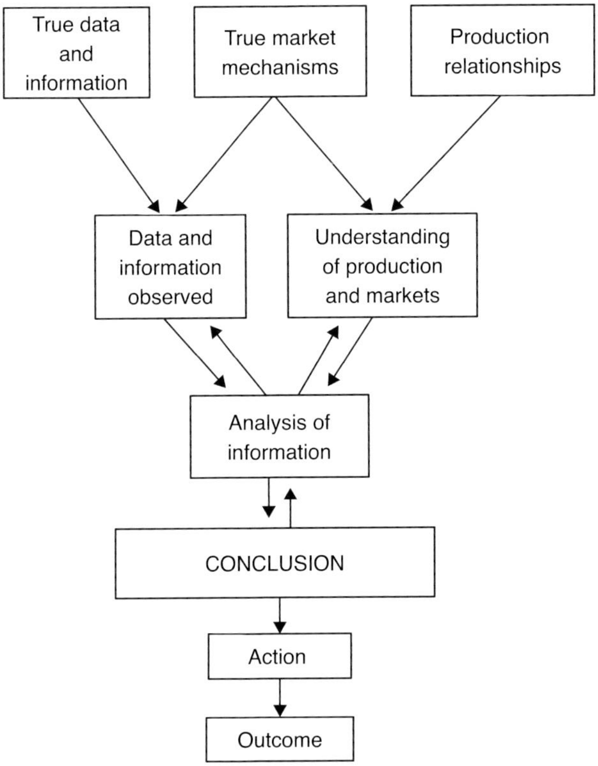

Fig. 6.1. Decision process components as sources of potential bias.

the lack of it in a consistent sense, will cause problems. Thus, someone who thinks the method they use when, say, organizing product packaging and delivery is appropriate, when in fact it is causing product damage, has a 'biased' procedure. These physical biases are not discussed here.

To categorize potential biases it is useful to think of the decision steps and biases in each step. Figure 6.1 lists the steps. The diagram reflects that information and data exist to be observed and noted. Similarly, the world around contains markets of various kinds which determine the factors that pass to and from the business entity – for example, prices of the products. Finally, the interaction of inputs and outputs is governed by a set of physical and, perhaps, biological relationships such as sterilization for preventing damaging organisms. The manager observes all this and decides what the values of the important information and data are, including market information. Whether these observations provide the relevant variables, and an accurate estimate and understanding of them, is another matter. This is reflected in the next row of boxes. The brain-held information is then exposed to the manager's cognitive processing to finally come up with a conclusion leading to action and outcome.

There are also feedback loops in that, for example, if the conclusion reached appears unusual, the manager may resample the information to come up with, say, a better estimate of the product price. Similarly, following the observation of the outcome, various assumptions and information might be altered.

Considering this whole process, biases can be grouped according to the boxes in Fig. 6.1. Thus, we have:

- observational biases;
- forecasting biases;
- decision process biases;
- implementation biases; and

- general biases that might impact on several areas.

In addition, as a manager works with people in the process of creating decisions, and their implementation, another bias category can be called:

- people dealing bias.

This allows for inappropriate relationships that lead to consistently less-than-optimal outcomes. Strictly this bias could be put under the 'implementation' label in part, and also under the 'observation' label.

The discussion that follows will consider each of the bias areas in turn, and conclude with comments about stress, before finally covering suggestions on observing and correcting biases remembering that a bias has been defined as an error which is made consistently, and the decision maker, usually, is unaware of its existence.

Areas of Bias

Observation

Most managers rely on their memory for a significant proportion of their decision information and so accuracy depends very much on picking up the true data as it is observed, and then storing it in such a way that it is correct when retrieved. The trouble is that memory imperfections (bias) are common. For example, the context in which something is observed affects what is stored and remembered. If, for example, a manager attends a meeting at another business and is shown an obviously excellent production-line system together with the output data that have resulted from a particular machine set-up application, it is likely this information will be easily retrievable. In contrast, if the manager reads about the same situation in a back corner of a newspaper, or hears a radio report, the information is less likely to be observed for successful future retrieval. In that most managers are what can be called 'concrete', or kinaesthetic, learners, seeing and doing tends to be a more successful way of observing. Thus, a *manager's learning approach can lead to biases.*

Then there is the *halo effect*. If something good is observed about an item, it is often assumed that all other characteristics of the item are also good. For example, if a particular type of machine has a high rate of output, the halo effect has the manager also assuming that similar machines also have efficient output. Clearly, these two parameters may well be divergent.

Another common problem is called the *anchoring effect*. This occurs where the introductory information provided influences the conclusion the manager makes on the value of a variable. For example, it has been found that people will come to different conclusions on the monetary value of an asset to be sold depending on the preliminary information provided or observed. So, if you said a small business sold for US$1,000,000, 3 years ago, you often get a different estimate of today's value if in contrast, say, you are told the valuer noted the business sold 4 years ago for US$900,000. A manager must be aware of this common bias and adjust accordingly.

The *framing impact* is similar to the 'anchoring' effect. This refers to the impact of the information, or its particular deliverer, surrounding the core information. Thus, a manager might come to a different conclusion, if, say, his spouse reads out information compared with the same information being read out by a professional consultant. The same sort of impact can occur if the material is presented in a professional glossy form relative to, say, a simple note even though the core information is identical.

Selective abstraction (or sometimes called 'selective perception') is another common bias. This refers to remembering only the data that suits you (for whatever reason). Thus, for example, a manager might only remember the output from a bread factory occurring when the machines operate perfectly, so that when budgeting expected income the 'average' output used is much higher than the true average. This is something that most of us do in that you remember the good things with the bad being shut out. Perhaps this is a survival attribute in that people tend to push the bad to the back of their mind so that they are positive about the future, whereas a person who has had a long run of bad luck might start to wonder if it is all worthwhile. Similarly, how often have you selected information that suits your feelings and so ignore the data that perhaps leads to a different conclusion? Thus, for example, if a manager really enjoys dealing with a favourite product/s she may well put the lowest possible output levels in her/his budgets when comparing the product profitability of other product type returns. The conclusion, clearly, reinforces her/his personal wants. This might not be a bad thing provided it is recognized that excluding products is in fact forgoing profit and this is a cost of sticking to this particular product.

The old saying goes 'one swallow does not make a spring', yet one good outcome often provides a manager with the conclusion that good outcomes will always occur. This is referred to as the *sample-size* effect. For objectivity a manager should adjust his estimates of outputs and costs according to the sample size on which the estimate is based. This bias is particularly relevant for managers starting in a new area or product type.

Finally, while the manager may not have any reason to make biased estimates, s/he may simply be using a *flawed procedure*, or is insufficiently trained to get the estimate correct, or reasonably so. For example, many managers will use an intuitive estimate of the market size for a particular product, or perhaps take a guess at the labour productivity in a grading machine. Similarly, they may use an eye estimate to judge the live weight of fish produced in a growth pond. Many such examples exist. However, the manager may be consistently underestimating or overestimating the true values. Experiments have shown that with proper training a manager can be both accurate and consistent in estimating most business factors. Often this training requires tests and feedback until accuracy occurs. In the case of growth pond output, this would involve making an estimate, then taking a sample before weighing it. This is repeated from time to time until constant accuracy occurs. Similar procedures would be used for most production output estimates.

It must also be remembered that the manager must 'observe' the production unit's objectives correctly. The objectives, of course, serve as the measuring stick when deciding which alternative is to be followed. This assessment must account for the business family's objectives as well as those of any other owners. Not all managers, of course, see it this way in that they believe it is their own personal objectives which should take precedence. This situation is fraught with potential biases; thus, *bias in concluding on the appropriate objectives* must be guarded against.

In summary, in the general area of observation, recognized biases are:

- flawed learning approach;
- the 'halo effect';
- the 'anchoring effect';
- the 'framing impact';
- selective abstraction;
- sample-size effect;
- flawed procedures; and
- misleading objectives.

No doubt, others also exist.

Forecasting

Having rightly or wrongly observed data and information, this must be turned into useful managerial material. This invariably involves forecasting future situations resulting from one or more decisions. Thus, a manager needs to forecast, say, the market size relationship with the prices expected, the likely supply of experienced labour, the costs of inputs at different levels, and so on. Estimates must clearly be available for all costs and returns for all inputs and outputs with information on uncertainty and risks. No matter which type, and part, of a business is considered, decision making involves forecasting in some form. Failure to get this correct on a consistent basis gives rise to bias and incorrect decisions.

Common biases in this area include:

- 'dichotomous' thinking;
- technology misunderstandings;
- risk estimate failures;
- giving credence to unevidenced guesses;
- 'availability and recency' effects; and
- generalization effects.

Dichotomous thinking involves a manager assuming that an outcome will be either good or bad. In reality most outcomes in the uncertain world of production follow a continuous curve from the good extreme to the bad at the opposite of the distribution. In assessing an option, just taking the good and the bad possibilities may distort the result because the bad may overbear the good, whereas there will be many possible 'good' outcomes that may produce the opposite impact.

The main reason for following the dichotomous approach is its ease of estimation. It is possible to mentally calculate the result of both a good and bad outcome and weigh them up. To do more than this requires pencil and paper using chance (probability) estimates possibly with computer help.

Once you start thinking about chances, or probabilities, a manager's understanding of risk and risk calculations comes into the picture. Managers tend to produce imperfect estimates of the probability of the various possible outcomes (*risk estimate failures*). Probability distributions are a further complication to the manager and are seldom allowed for in the mental, or otherwise, sums. The notion of an 'expected outcome' (the sum of the possible outcomes each weighted by its probability of occurrence) is difficult for managers to grasp. They often think in terms of single outcomes as this is less challenging than the idea of a distribution from which only one event will occur in any one year or trial.

How do you accept the idea that the decision made turns out to be incorrect in hindsight whereas in reality it was indeed the correct decision, it is just that chance created one of the less likely set of circumstances? Perfect prediction is not possible in a random world, well not yet anyway. The overall outcome is that managers do tend to struggle with risk and uncertainty and tend to work with single-valued estimates; these might be called 'certainty equivalents'.

If a manager is unaware of the *true technical relationships* determining outcomes from an input set, then clearly the forecasts of outcomes will be incorrect. Any manager who consistently uses an incorrect mental production function model, for example, will end up with biased decisions. Deciding on appropriate machine configurations, or perhaps the mix of markets to supply, requires knowledge of the response relationships, market/price information and the like. If this is several per cent above, or below, the real relationship, or the turning points are wrong, the production decisions will be consistently wrong (biased). In reality it is doubtful whether any manager has the true production relationships stored away as this would require extensive trial work from her or his particular business, but s/he should at least have a general idea using the results of years of observations on her or his business, and similar businesses, and from research establishments that might be located in a similar situation.

How often have you made a guess? Managers are inclined to make guesses when forecasting as this is easier than searching out what information might exist. If a manager is inclined to do this on a regular basis, it is probable that the forecasts will be incorrect and, therefore, biased. The manager's personality might be such that making guesses is part of what might be termed a *lazy bias*. This does not mean some manager's guesses will always be incorrect. Indeed, some managers may take in more information than appears to be the case and their 'guess' is actually based on considerable information and frequently relatively correct. However, the degree of guessing combined with a manager's inappropriate intuitive observation powers gives rise to some managers being consistently wrong. If the degree of incorrectness is sufficient, they will, of course, end up struggling financially.

The last well-documented forecasting biases are the *availability and recency effects*. There is a tendency to use data that is readily available simply because of its ease of access. Its accuracy is not checked and so its use may lead to errors. This could be called a part of the *lazy bias*. Similarly, managers tend to remember and use data recently acquired in contrast to adding the recent information to the store of past information in order to create a longer-run set. Clearly this must be guarded against.

Overall, it is prudent for managers to be constantly comparing the results of their forecasts with what occurs and adjusting their procedures and mental assumptions and pictures in the light of the feedback. If this is successful, biased forecasting will be kept to a minimum.

People relationships

Managing a business involves constantly dealing with people, including the business family members who may have both a managerial and physical input, either full- or part-time, through to lawyers and accountants who deal with the ownership, resource consent, employment aspects, through to taxation reports. In between are employees, competing businesses and other local colleagues, commercial agents of various kinds, professional consultants, bank managers and contractors. All these people are the manager's interfaces to planning, getting the jobs done and meeting all the regulations. Good relationships are critical to success.

Stories of which businesses to avoid when looking for work abound in some districts. A business is often an intimate place of work in that the manager and employees are in constant contact and so if their personalities are opposed it becomes a difficult working environment. Similarly, successfully dealing with contractors who have many clients wanting their job completed immediately is crucial to success. While less frequent, ensuring the contacts with lawyers, accountants, consultants, bank people and the like are all appropriate and give rise to the outcomes desired is also important.

Unfortunately some people, even with the best of intentions, manage to upset others so that their dealings are less than successful in completing the job as required. Others are downright rude and cantankerous, characteristics that will not build a supportive and achieving team. Anyone who lacks all the people skills required can be said to have a bias in that they consistently end up with unsuccessful dealings with people relative to other managers who have the knack of getting what they want achieved.

The whole field of human resources has spawned many books on how to succeed in this area, so the skill is in recognizing that there is a bias and setting about solving the problem. Of course, this may be difficult where large personality changes are required. However, at least recognizing that there is a problem is a significant first step.

Also relevant is the manager's ability in assessing potential employees (*people bias*). It is relatively common to misjudge how a person will react in a range of business situations. For example, someone who has experienced good luck might be deemed clever, a person with a dominant personality tends to be believed despite the facts, people try to act according to their assessment of what they perceive the manager would want, and, similarly, potential employees tend to follow what is regarded as normal even though their beliefs might be different. When taking on staff, and indeed when judging people who interact with a manager, the manager must be aware of all these tendencies and make a judgement of the person's true personality.

Decision principles

Assuming that the information being used is correct (which may not be the case), it is then necessary to use it logically to provide decisions and consequent action that achieve the objectives. Thus, any thought process that is illogical creates a bias. In general, without a good understanding of the production economics decision rules outlined in an earlier chapter (Chapter 4), there is likely to be, on average, constant decision bias. Of course, one of the most difficult areas is the handling of risk and uncertainty. This requires the manager to be fully aware of the risk attitude which is part and parcel of the unit's objective function. This must then be built into making the decisions.

Within this 'decision principle' area, there are clearly many possibilities for bias including:

- incorrectly allowing for the risk attitude within a production economics context;
- using averages instead of marginal effects;
- ignoring the time value of money;
- using an inappropriate mix of inputs which do not produce least-cost production;
- ignoring the opportunity cost of unpaid inputs; and so on.

Overcoming these biases requires a good understanding of all the decision rules together with a careful analysis of the decisions made. Due to risk, uncertainty and time impacts, it is not a simple matter to come up with the right decision to compare with what was actually decided and therefore test for bias. Given the constantly changing technology and markets, years of comparable records will seldom exist.

Besides decision rule biases a number of other biases have been researched and recognized. These include:

- chunking errors;
- overconfidence;
- automation bias;
- 'lazy conclusion' (impatience);
- self-esteem errors;
- 'good follows the bad' syndrome; and
- 'market upturn hopes'.

A business can include a complicated production and marketing system which is difficult to wholly comprehend and mentally picture. Consequently it is common to break a problem into components called 'chunks'. The trouble is that many parts of a business are interrelated, thus making separating out independent components difficult.

A manager, for example, decides to compare all the products he might produce and so calculates, perhaps, the per-machine-unit return, or perhaps, the dollar investment return, by deducting the variable costs from the gross income. S/he might decide that they should, say, produce nothing but product one as it has the highest profit. The 'chunking' approach has been used in that each product creates a chunk to be analysed for comparisons.

In reality, it might be impossible to produce nothing but product one as the available equipment and labour cannot cope when it all comes on stream, thus a proportion of the output might well be left too long and will deteriorate both in quantity and quality.

If the 'whole business' aspects had been considered, it might be clear that the manager should stagger production runs for a range of products, thus spreading the labour and machine load to produce a practical solution. Of course, allowing for the 'whole business' would make the profit estimates more accurate. Perhaps lower gross returns would reduce the profit estimate for the whole business when taking producing just product one into account creates bottlenecks leading to production difficulties.

Perhaps different 'chunks' should have been analysed. A subset of products might compete and form an independent unit that can be compared against other independent subsets and, therefore, can be considered separately in their own right.

Another common bias is 'overconfidence'. It is a human trait to imagine that we are more competent than actuality, though some people suffer from the opposite and always downplay themselves. This can be just as much a problem. However, if a manager is consistently overconfident this may not impact on the choice of what, and how, to produce, provided the options stay in their right order, and it is similar for under-confidence. However, despite the correct ordering of alternatives, problems can arise as the overconfidence may lead to bad errors and disastrous outcomes. For example, confidently thinking that their skills will get them through a bad winter may mean a labour manager will get very poor production, and even labour shortages, resulting from this overconfidence.

Related to overconfidence is 'automation bias'. There is evidence that humans tend to uncritically accept computer-generated solutions and information. This confidence can be totally misplaced in some computer systems.

A manager's self-esteem can similarly lead to bias. A low esteem can lead, for example, to low market returns as the manager accepts the low offers made for his products. It is easier to accept what is offered in contrast to checking the market and bargaining for an appropriate price. A low self-esteem might also lead to a very conservative approach to decisions and so opportunities are missed which, while having some risk, on average are likely to provide a higher return. Self-esteem can also influence people's relationships.

Impatience is another source of bias. Searching out all relevant information, and its careful consideration, is essential for good decisions. Yet how many people are somewhat 'slapdash' in doing this for they lack the patience to conscientiously and methodically follow through all aspects. Impatience will relate to a manager's personality, so it is important for the manager to recognize this and work hard at curbing the desire to proceed before the decision research has been completed.

Impatience also impacts on day-to-day management. For example, a manager may wish to send a new product to the market before a well-thought-out packaging and marketing programme is organized and set in motion given her or his excitement at getting this far. Impatience might also be called 'lazy' decision making, though the root cause is probably different with personality and experience factors involved.

Overconfidence, an improper self-esteem and impatience can also all lead to the manager misinterpreting the business unit's objectives. This is particularly the case for self-esteem. As noted, getting the objective function right is crucial to success as without a correct yardstick very little will be right.

Finally, two other commonly observed biases need mentioning. Frequently you will hear a manager comment that he will persist with a project although past conditions have produced bad outcomes, as this luck must soon turn to give good outcomes ('good follows the bad' syndrome). For some reason, if poor conditions have occurred managers assume that the probability of better conditions must have increased. In the case of weather, for example, monthly events are likely to be independent of previous events. Bad does not create good, though in the longer run, in the case of the weather, averages do tend to remain constant so eventually the 'bad' will be compensated.

Similarly, where a manager has invested in a project, say a newly bred and propagated intensive flower crop, where there has been a string of market signals suggesting the new flower will be unprofitable relative to the alternatives, there is a tendency to incorrectly continue with the new 'wowing' colour and form as it is expected there must be an upturn in the yields and markets ('market upturn hopes'). Unless there is very good evidence to support this, it may be better to cut the losses and move into another enterprise. This is all part of the syndrome 'after bad the good must come'. While this might be a good optimistic view to bring to life, it must be tempered with the real probabilities. This bias is part of the stubbornness of man. 'I thought of a good idea ... it must come right so I can recoup my losses.'

Implementation

Many of the biases mentioned also apply to the implementation phase of any decision. Thus, overconfidence can lead to consistent bias, as can probability estimate failures, and an irrational faith in continuing with a project despite the poor outcomes. Probably two of the main biases are the:

- incorrect prioritization of the tasks involved in implementation; and
- the timeliness of actions.

Starting disease control in an extensive glasshouse-based flower production factory well after the indications that it may be a problem will cause constant difficulties. Usually early recognition and handling of a problem gets the best response, so a manager who constantly fails to notice the early signs will suffer relatively low-quality outputs. This is both an observation and timely action problem. Even if an emerging problem is noticed, if nothing is done to rectify it, the result will still be poor. Another example might be the arrival at market of quality-degraded products due to, perhaps, transport and environmental control problems.

This leads to the other main bias in implementation:

- a failure to constantly control and review plans.

In the ever-changing and dynamic production world, conditions are usually different to what the norm might be, and similarly the prices and costs are seldom as anticipated. The logical response to these changed circumstances is to review the progress of the project to see if a change in the plans is appropriate. In some cases, it will be optimal to continue with the original plans, but in many cases a change will be optimal.

Thus, for example, it may now pay to purchase double the supply of raw materials originally planned when both the costs, and the anticipated availability, are expected to be less than originally expected. The manager must look at the situation from the point of view of the current condition of the project, and what the anticipated responses and prices are. It is from this constantly changing base that the new optimal decisions must be calculated. Sometimes these will be the same as before, and in other circumstances changes will be optimal.

It is the failure of some managers to constantly update plans that gives rise to a bias. They must learn to accept that the decisions and plans made in the past are indeed history and should be forgotten because you cannot change the past. It is the future that must be focused on through a stocktake of the current situation and the calculation of the next best move.

General

Finally, there is one other common bias that concludes the list of the best known. This is:

- an aversion to change.

This attitude is probably related to both a manager's personality and past experiences involving failed initiatives. Some people just do not like changing their old habits to operate optimally for the current conditions. They have been introduced to a particular business system, probably from a young age, and feel comfortable operating the particular system. They tend to continue with the same set-up no matter what happens and hope that it will bring them through to at least a reasonable living. Of course, if the conditions change too much, they will end up losing their business.

There is something about such people that makes them feel very uncomfortable with change. Probably most humans have a component of change discomfort for it involves venturing into the unknown, and, therefore, creates challenges which some think they may not cope with. Clearly, sticking to just the known will lead to constant bias in most modern, ever-changing business environments. Managers with this aversion to change clearly need support and encouragement from their families and friends as well as from the professionals who deal with them. Seeing others successfully making changes is also useful, and thus the value of business colleagues getting together to compare and discuss systems, particularly when not in direct competition in a smaller market.

Stress and Decisions

Introduction

As noted earlier, stress can give rise to biased decision making. A review of the literature makes it clear that stress is a very real phenomenon in rural business and production, as is depression. In extreme cases, suicides result. In France, for example, Fernet *et al.* (2016) found and discussed burnout and psychological issues in SMEs, and research in Britain concluded that 6% of the rural managers had 'clinically relevant psychiatric morbidity' (Thomas *et al.*, 2003). This was similar to the general population, though a greater proportion tends towards suicide. Of course, while these extreme cases are serious and need timely intervention, the total cost of stress that is less extreme is significant for there is constant bias as a result (Schreibauer *et al.*, 2020).

Some managers can be classed as 'relaxed' in that no matter what happens they simply get on with the job in an efficient and logical (unbiased) way. Others get quite stressed when situations are not as expected. Undoubtedly their personality is a major factor in how they cope with stress, as is their past environment. Many people will have come into contact with people who find problems in whatever happens due to the influence of people they experienced as youngsters. In many cases a really difficult situation does not in fact exist, it only appears this way to the stressed manager. Thus, a combination of genes and environmental experiences moulds how we will react in what some would call difficult circumstances.

Difficult circumstances can come about due to a range of factors, and some people react to some, but not others. The following is a list of possibilities:

- Weather impacts, particularly extreme and prolonged situations that seem to be increasing under global warming.
- Financial impacts such as the price dropping considerably for a product that has involved major investment, and such as very high interest rates with high debt.
- Technological downturns and problems such as a failure in a software product that is key to the business (perhaps an attack from some hackers).
- Overbearing regulations that require complicated paperwork and reporting (health and safety requirements, perhaps).
- Relationship problems, both on- and off-business. It is difficult to escape situations on a family business.

Fig. 6.2. Stress takes many forms. Overcoming such disasters takes support and strength of character.

- Isolation and the working environment. Often managers spend very long hours by themselves, particularly in bottleneck times such as public holiday periods.
- Environmental pressures. Increasingly there is pressure from the general community over protecting the environment and complying with regulations in difficult situations.

It is a fact that dealing with the elements, world markets and people relationships including councils and governments in a small enterprise can be difficult, especially when a manager's life's assets are tied up in the business. It can be a fraught operation. Every manager must be aware of potential stress situations and guard against the impacts of this stress (Fig. 6.2).

Combating stress

To combat stress the person must be clear what causes the stress, and, of course, in the first instance understand exactly what stress really is and when it is occurring. Stress will be created by any event, or anticipation of an event, that is seen as endangering the person's financial, physical or psychological well-being. Thus, for example, when it is discovered that income will be much less than what was expected, and planned for, a stress reaction in the business manager might well be created.

In this case, it is probably quite clear what has caused the anxiety. However, what one person finds stressful might not impact on another in the same way. Where stress does occur, recognizing the cause is the first step in combating the debilitating impacts. For a full description of stress and its alleviation see, for example, Atkinson *et al.* (1999).

There are two general approaches to deal with stress:

- the first approach is to do something about the problem that is giving rise to the stress; and
- the second approach is to do something about the mental reaction to the situation, and, therefore, reduce the impact of the 'stress' and the associated anxiety.

Combinations of the two approaches may also be relevant.

Modifying the situation can take several forms. Maybe the stress is caused by worries about meeting production targets under a variable labour supply and climate. If the potential shortfalls can be overcome then the stress is removed. Possibly investing in storage areas and putting aside an inventory when production is working well, where this is feasible, may both remove the stress and increase profit. Or perhaps competing businesses might be contracting to produce stocks which might be built up to cater for the variable supplies. Or perhaps a positive bank balance is kept and maintained to allow open market purchase when required. The advantage of cash reserves is that they can be used in any potentially stressful situations and, thus, remove many problems that might cause stress.

Other options for removing the problem include *learning new skills* (perhaps forward trading to cover exchange rate worries), *using contracts* to set the price received, and similarly contracts for physical tasks might help in 'bottleneck' situations.

Then there is the opportunity to *alter the mindset that creates the stress*. Perhaps, for example, the manager might review his objectives and consequently no longer worry about having, say, the highest market penetration in the district.

As noted, the second side of the coin is examining the manager's mental reaction to the situations causing stress in an attempt to remove anxiety. The objective is to block, or change, the emotions that overwhelm the manager and lead to irrational decision making.

One of the first things is for a manager to *keep records of the situations that cause stress*. Some of these might seem trivial, but a full record helps to make it clear where the problems lie. Until this is sorted out, it is difficult to look for remedies. Perhaps, for example, a manager is stressed over how many person hours the business is losing due to sick leave and dilatory work hours. The solution might be to have an expert check the relationship status between management and the work force, or certainly keep detailed records of both attendance and productivity. These might well both fix the problem and set the manager's mind at rest. Alternatively, maybe the manager has to *change his mental attitude* to the problem in that, in the example given, the amount being lost is in fact normal and so the stress reaction was a needless self-creation.

One approach that many take is *blocking out the* problem. Sometimes a manager will do this through alcohol, or even drugs, or simply through denial. None of these approaches is likely to help in the long run, though blocking out a problem until the last minute before action must be taken can help. Why worry about a situation that is not yet here, and currently requires no action?

Sometimes blocking, or denial, subconsciously leads to unidentifiable concerns, even perhaps impacting a manager's health status. Such a situation, if noted, probably needs professional help. On a more positive note, *exercise can help* modify the impacts of stress, particularly recreational exercise which provides further distraction from the problem.

A generally healthier approach is to *share the problem* with family, friends and/or neighbours, and also professionals (Beckmann *et al.*, 2023) such as a consultant. There is evidence that 'a problem shared is a problem halved' leading to less stress and rational solutions. The *act of talking about a problem helps sort out a logical solution* and once this is achieved the stress may well abate. Clearly this approach to finding a solution is by far the best strategy if it does reduce the stress for it may well also create a rational outcome. Some of the other strategies may not.

Given any stressful situation it is also important to think it through for it *might not have been avoidable*, and there may be nothing that can be done. Something like an unforeseen market collapse clearly is not under the control of the manager, and so it is important for the manager to think this through and realize that it was beyond his control and so the stress felt, while understandable, is non-productive.

Another approach that may help in some circumstances is '*displacement*'. This refers to altering, or replacing, a motive that has caused the stress. For example, a manager may have an unrealistic expectation over the profit that is possible given the skill levels and type of resources. Perhaps the *expectation can be lowered* and in its place an expectation of greater leisure time created. This might involve an objective to lower the golf handicap!

Effectively this strategy is one of changing the manager's aspirations to a level that does not create overburdening stress. Another term sometimes used where a manager's *beliefs are different from actual behaviour* is '*cognitive dissonance*'. To remove the stress the belief needs adjusting to be more realistic.

On a more positive note, there is evidence that creating positive illusions can help stress coping (Taylor and Armor, 1996). The manager needs to visualize the stress-creating situation before it occurs, and then talk himself into being relaxed about the situation through bolstering his self-image and perceptions of control. Effectively the manager is *simulating the potential situation and working out how to handle it* under the assumption that he has all the skills necessary, and it will all be a smooth operation without any stressful outcomes. Certainly in the sporting arena there is significant evidence that visualization of success does in fact assist success occurring.

Another positive approach is for the manager to *learn* relaxation *techniques*, and to ensure he has plenty of aerobic exercise, both of which can reduce the impact of potential stress. A relaxed person is in a state opposite to that of a stressed person. There are many available courses and books on relaxation techniques (other than counting sheep, or some other object, when trying to get to sleep!). Most techniques involve working on relaxing the muscles, from toe to head in that order, ending up with a consequent mental relaxation.

Finally, the manager has the choice to *avoid situations that cause stress*. In effect, this might be achieved through some of the strategies already mentioned. For example, use of forward contracts might well remove stress caused by variable income, reducing the number of labour units employed, and so on. There are myriads of strategies that can be used to dampen the stressful situations, each one of which should be examined in the search for a better system.

It should be noted that dangerous stress situations may need professional psychiatric help. Anyone involved with managers in an advisory/consultative capacity must be very careful to seek such assistance whenever there is real concern over the health of a manager. The manager's family will more than likely be similarly seeking assistance, though sometimes it is easier for an outsider to organize this. If the manager is in denial, the situation is that much more difficult.

It should also be remembered that within the rural population there is a continuum of stress reactions and feelings right from the extreme level, referred to above, to the very mild which has no harmful bias impacts.

In summary, the main techniques discussed here for reducing stress involve:

- keeping records to first identify the causes of the stress;
- altering the system giving rise to the stress;
- learning new skills allowing the use of stress-reducing techniques or approaches;
- use of risk-reducing techniques such as forward contracts, diversification and the like;
- changing the objectives/attitudes (called displacement);
- sharing the problem with friends/relatives/professionals;
- positive illusions;
- relaxation techniques; and
- avoidance of the stress-creating situations.

The Case Managers' Thoughts on Bias and Stress

Margrave's case

Margrave certainly experienced some stressful times. One example involved a rival for a position where the competitor was a local person. The family received threatening phone calls, making life very difficult for a time. But in the end, the people skills Margrave exhibited meant he eventually developed a good rapport with the person and often chatted to him about business issues. Perhaps Margrave had many intuitive skills which automatically came into play. He recounts the story of buying a holiday house, or rather the attempted purchase of the house. It was in an idyllic spot near the sea with a facility for Margrave to store one of his few off-business passions, a trailer yacht. The night before the auction Margrave had trouble sleeping and

Fig. 6.3. Ensuring machines are fully used and producing quality output can be a major source of stress, particularly where skilled labour is difficult to obtain and supply chains are problematic. (Courtesy of Fahad Faisal, Wikimedia Commons.)

woke several times. He commented: 'One thing I never did have was trouble sleeping.' In the end, he decided against attending the auction and later discovered that the historic house was on a spot that had once seen a massacre of the local indigenous people who were ransacked by a visiting chief and his warriors. Can this intuitive response be logically explained?

Margrave believes he has a bias over being too kind to people and does not dismiss or reprimand staff when perhaps he should, but despite this bias he certainly seemed to be successful with labour relationships. Margrave also believes he is perhaps insufficiently anxious on occasions when he should be worrying and sorting out an issue. However, he always does an enormous amount of homework and so there is seldom a need to worry for he has confidence in his conclusions. In the end, he walks away from decisions and situations he is not comfortable with. Is this a bias?

In his desire to have high production, Margrave believes his output is not as high as it should be given the resources and inputs used and that if his labour force was more skilled the output would be greater. He noted: 'I have rosy tinted spectacles' and 'should learn to 'back off''. Margrave is the total optimist and thinks that the next spring will always be one of the best, with markets providing upturns (particularly with respect to wool output use. See Fig. 6.3). This 'does get me into trouble'. To counteract this optimism he keeps a large photo of his business in the worst year for producing quality wool pinned to his office wall.

Profit was always at the back of Margrave's mind and forecasted outcomes influenced decisions. However, on a daily basis, 'financial aspects don't influence daily management and I probably don't do enough cash flow forecasts ... however, the situation has never got bad enough to create depression'. Margrave always discussed most situations with his wife and consultant and so this sharing may well have helped prevent financial stress. And to further help his outlook, both mental and physical, Margrave always moves to ensure his office was close to successful parts of the business which 'helped morale enormously'.

The comment about cash flow budgets is also reflected in his attitude to records. He notes: 'I enjoy the technical aspects of business, but I certainly don't like keeping records and monitoring. Records ... what for?' However, Margrave always weighs inputs and outputs assessing their condition, and carefully notes bank balances and input inventories enabling 'instant access'. This reflects his excellent memory and his confidence in his ability to assess all relevant facts without the stress of keeping endless records.

Now Hank's views

Hank also recognizes that he does have some biases, and certainly experiences stress at times. It would be surprising if good managers never experienced stress for part of being excellent is being human with the associated emotions and drives.

Hank comments: 'I'm just no good at formally monitoring what happens around the businesses ... I find it boring, and what's more my monitoring skills are terrible. For example, no matter how hard I try I'm useless at judging potential employees, and equally no good at accurately estimating input costs for future years, let alone output price trends. What's more, I'm not only well out, but my estimates are inconsistent.' Thus, Hank relies on staff to carry out these monitoring tasks. Hank freely comments that he finds the planning and execution side of management challenging, fascinating and enjoyable. This is where he spends heaps of time. What Hank recognizes is his areas of bias and weakness, and so adjusts accordingly. This is clearly very healthy.

Hank also recognizes that he has 'rosy tinted spectacles' when it comes to estimating input needs, with understatement being common. Hank comments: 'I always underestimate what is likely to be required to meet upcoming contracts.' However, he updates his cash flows every 2 weeks so that the underestimates seldom have time to lead to major problems. Of course, these are self-recognized biases which are compensated for. The real problem is the existence of other biases which are not corrected. This is where a third party can help, or perhaps family members, or even trusted employees.

Perhaps rationally, Hank also believes he is biased against intensive high-input-level systems that require specialist machines and inputs with a need to ensure few breakdowns. He maintains there is insufficient extra profit for the risks involved in this intensive production. Small decreases in product prices leave you teetering on losses, and with such intensity, a small error or problem causes havoc to the profit. Perhaps he is right.

This attitude, which Hank calls a bias, is perhaps related to his attitude to risk and stress. He just does not want to live on such a knife edge. Hank recounts how he does get stressed at times from the 'nasty volatile markets'. He can remember one season when the supply chains for many inputs started to have significant problems. He spent many nights ringing potential suppliers and transport people on the other side of the world. It became a vicious circle with middle-of-the-night panic attacks. The situation had got the better of his rationality and emotions. Help from others finally led to thinking through the problem and the return of rationality. 'It was pretty scary.' But it was also a learning curve about putting circumstances into perspective and acting appropriately. Hank would not be the first manager to face such problems, particularly exacerbated by high levels of debt.

Concluding Comments and Bias Reduction

Most managers exhibit both some form of biased decision making and stress. The latter, however, may not impact on efficient decision making, but where it does, remedial action is desirable, just as it is for non-stress-related biases.

The main difficulty is recognizing the bias that exists because once the bias is recognized and annotated, correcting the problem is likely to at least a certain extent. Correcting overbearing stress, however, may be more difficult.

There are three approaches to recognizing bias:

- The manager himself needs to be constantly aware of the possibility of bias, so self-examination is the first approach.
- The second is to employ other people to consider the possible biases and provide suggestions as to where improvements might be possible.

For the manager to employ other people requires a certain amount of humbleness and a willingness to accept an expression of poor management as well as a willingness to change.

- Third, it is conceivable that formal tests might indicate areas of bias.

Currently such tests are not generally available, but, for example, it should be possible to present well-described and well-illustrated decision problems for which the manager is asked to provide the answer. If the simulated decision situations are reasonably realistic, it should be clear if bias exists. Of course, some managers will not manage written tests as it can be hard to relate paper situations to reality. Currently, however, bias recognition must be human based.

For self-assessment, it is useful for the manager to have benchmark data on the performance that might be expected. This should ideally be from other businesses in the same environment and start with profit information. Detailed benchmark data can go down to, for example, per-labour-unit output for different input levels. Such data needs to be carefully collated and allow for seasonal variations through having data for as many years as possible.

One year's data is unlikely to be useful for random impacts can influence the comparisons. It must also be remembered that the data might not reflect the ideal for while some managers might perform better in some areas, they may well be worse than the case manager in others. Thus, while a comparison is essential, it must be used intelligently.

Any comparison process must take cognizance of the business's objectives because profit and cost might not be the main priority, though physical input/output ratios are likely to be universal in that few would not want maximum output for given inputs. Again, however, production economics principles must be allowed for in that, for example, maximum output per technical unit (e.g. machine) is unlikely to be optimal.

Where outside people, or even family and colleagues, are asked to help, they will similarly have benchmarks in mind to allow some sort of comparison and conclusion. Some of these benchmarks will be related not only to their experience of other businesses, but also of other managers. This is the additional judgement other people can bring to the assessment relative to a self-assessment. While such judgements will be subjective, an experienced person will certainly have ideas of where the bias occurs. One obvious area is how the manager deals with other people. While in the end there must be opinions involved, a good manager will welcome constructive ideas as food for thought and possible action.

Little can be said about the formal use of tests as they are yet to be developed. But examining how professional consultants assess a manager may lead to ideas of what might be included in such tests. In that personality and management style are more than likely related to biases, these tests will be of assistance in assessing a manager.

For example, a judgement on a manager's anxiety personality trait can be particularly helpful in assessing stress levels. Other traits will similarly influence outcomes, another example being the manager's attitude to new ideas (openness). As bias is related to personality, style, experience and training, the previous chapter (Chapter 5) contains a more detailed discussion about how these factors might be altered to give greater management skill. After all, this is the main objective in considering a manager and his attributes.

Suffice to comment at this stage that probably one of the most important skills in noting, and working on improving, biases is the ability to be self-critical. By taking a critical approach is not meant being negative about everything; but being objectively analytical in looking for biases and working on their correction. Furthermore, of course, a critical approach in this positive sense is the most important attribute for all areas of business.

When, for example, a commercial agent tells the manager about the large number of positive attributes the product he is selling has, each should be carefully examined to assess whether the benefit would be achieved on his business. Effectively, everything should be taken with a grain of salt until clearly proven.

As part of this critical approach, all decisions should be double-checked, and, where possible, discussed with others, for this process clearly outlines what is proposed and why, and, therefore, leads to both a self-review as well as others' critical input.

In this process it should be noted unfavourable situations, like extreme tiredness, can lead to biased decisions (Croskerry *et al.*, 2013).

In the end, change and improvement must come from within.

7 Intuition

Abstract

As 'intuition' is a widely used concept, this chapter fully explains its structure and improvement. Intuition as a decision skill improves with time (age) as experience provides many lessons. While the changes can be cognitively automatic, procedures which enhance change are discussed with the concept of 'pattern matching' introduced. An analysis of managers with extensive useful experience highlights the attributes excellent managers acquire. A hypothesized model of intuition is fully described and analysed. The results show the importance of all the constituent variables, with the most influential being 'technical knowledge' followed by 'knowledge of decision theory'. Reflection (self-critique) is also significant. A comparison of managers with high intuition relative to others also stresses the importance of the personality factors. To simplify the details of the initial analyses, another analysis of combined databases elucidates the contribution of three core variables: phenotype, training and experience. These summarize all the variables presented in the first model. Information on how to improve intuitive skill concludes the chapter.

Introduction

There is little doubt most SME decisions are made by managers using their intuition. The resulting conclusion is their mind's answer to resolving the issue to hand. This process might be almost instantaneous, or the mind might come to an answer after a little time and reflection. Whichever the case, the decision does not involve formal and recorded analysis.

It is thought by some that 95% of decision making is based on intuition (Croskerry *et al.*, 2013), whereas others might not put an exact figure on the proportion but are sure intuition is the dominant decision system (e.g. Ohlmer, 2001; Chaston, 2009; McCown *et al.*, 2012; Nuthall, 2012).

The discussion in Chapter 3 on the results of modelling managerial ability shows how important experience is as a factor in ability, though it must be acknowledged this was based on judging success on monetary returns only. Indeed, the data shows experience is about four times as important as all other factors such as management style and intelligence. This knowledge, however, begs a question for it might be asked what is the exact meaning of 'experience', and how can it be improved as a factor in managerial ability? This is an important human factor and is explored in this chapter. However, if other legitimate objectives are included in an analysis the importance of experience is not as dominant (see later discussions).

In reality, experience largely interprets into 'intuition'. Without experience decision makers do not make good decisions. For a manager who has never 'managed' before, and knows little about the specific business type, it would only be by chance if their decisions turned out to be reasonable. Experience is critical to success and leads to a manager's intuition provided the lessons available from each experience are noted and used for future decisions, hopefully after improvement following each repeat occasion. In assessing decision-making success, however, it must be remembered success is not necessarily high profit. What is important is that the manager achieves her or his objectives no matter what their form. As has been noted before, leisure and perhaps maintenance of the environment, for example, might be just as important as cash surpluses.

There are many definitions of intuition. One definition expresses that decisions made through intuition are 'reached with little apparent effort, and typically without consciousness awareness. They involve little or no conscious deliberation' (Hogarth, 2001, p. 14). Hogarth (2010) also regards intuition as a person's cultural capital though others call it part of a person's human

DOI: 10.1079/9781800627246.0007

capital. Dane and Pratt (2007, 2009) also talk about intuition as providing decisions without formal analysis which are largely non-conscious. Another term for intuition is 'tacit knowledge', which is regarded as the mind's store of decision rules and background information (Polanyi, 1966). Both the manager's intuition and tacit knowledge, which are likely one and the same thing, is certainly one of any production entity's resources just the same as land and capital. Social capital, being the sum of a manager's community relationships, is also important and a further human factor important in primary production management.

Intuition has several parts. Before the correct decision response can be produced, 'intuition' must (Eraut, 2000):

- first, observe and understand the current situation and expected future conditions ('reading the situation');
- second, mentally interpret and analyse the information; and
- finally, frame the conclusion to suit the current decision problem.

These steps are automatic. Indeed, what is called pattern matching (Dane and Pratt, 2009) is important because part of this process of automatically searching the mind's memory slots is to come up with a solution for the exact situation previously concluded and stored away.

The concept of intuition was introduced in Chapter 4, but this chapter expands on the concept through reviewing past work and outlining a set of research activities designed to discover the important components of intuition. The explanation covers the development of a model of intuition and its quantification to discover the relative importance of the variables hypothesized to contribute to a manager's intuition. This then leads on to a discussion on how this knowledge can be used to improve the adequacy of a manager's intuition as this must be a major goal of studying the human factors in management.

It needs to be clear that intuition is a learnable component of human capital. Only a fraction, if any, of a person's intuitive skill is part and parcel of their genotype as was noted when highlighting the skills of a novice manager above. This is in contrast to skills like avoiding danger and exhibiting high basic intelligence. Survival demands, for example, on automatically and rapidly removing your hand from boiling water. This is an example of Darwinian survival of the fittest. Intuition is not the same per se.

The Background to Intuition

It has been noted experience is a key factor in managerial ability. This leads into experience's modus operandi – that is, 'intuition'. In this sense it is important to consider the factors associated with valuable experience. The study on managerial ability reported in Chapter 3 provides an opportunity to compare managers with valuable experience relative to those not utilizing their experience to the full. This provides ideas on the important factors. Similarly, a review of the literature covering past studies on intuition provides further ideas and is highlighted in a following section.

The results of the modelling described in Chapter 3 allowed rating managers on their level of success in using their experience (Nuthall, 2013). This allowed comparisons. Managers with an experience percentage rating of greater than 80% were compared with those having a rating below 50%. The results are shown in Table 7.1. Only the factors that were statistically different between the groups are presented, and only those likely to enhance useful experience included.

While there are many factors to consider, the highlights clearly show managers with high experience are much better at achieving technical excellence (productive efficiency) compared with the others. You would expect this for one of the first things associated with high managerial ability is obtaining the maximum physical output possible with the resources available. Efficiency studies make this very clear just as it is obvious logic. And according to the data, being younger also helps!

It is also relevant that managers prepared to talk to others about their decision problems obtain high value 'experience'. Similarly, managers who write down ideas and budget comparisons tend to have greater success.

However, the data shows anxiety and worry are not good traits to have. Becoming less anxious

Table 7.1. Comparison of statistics of managers/businesses with useful experience rating percentage > 80% ('High Exp.'; 41 businesses) relative to managers/businesses with an experience rating < 50% ('Low Exp.'; 72 businesses). Total sample = 735 businesses.

	Mean value	
Managers with:	High 'Exp.'	Low 'Exp.'
PARENTAL INFLUENCES		
Attendance at rurally based primary school (1 = 'yes', 2 = 'no')	1.1	1.3
Encouraged to use imagination when young[a]	2.0	2.5
Encouraged to get on with people when young[a]	1.6	2.4
Listened to technical matters discussion when young[a]	2.8	3.3
Wanted info on decision reasons when young[a]	2.4	3.2
PROCESSES		
Discussions with others helpful[a]	2.0	2.5
Write and calculate monetary consequences[a]	2.2	2.9
Talking to others provides enthusiasm for ideas[a]	1.7	2.7
Improved my management of workers and contractors[a]	2.3	3.1
Plan formally with paper budgets and similar[a]	2.3	3.0
Ensure employees enjoy their jobs[a]	1.3	2.0
PERSONALITY and ATTITUDE FACTORS (BELIEFS/FACTORS)		
Easy to ring strangers for technical information[a]	2.4	3.6
Tolerate employee/contractor mistakes[a]	2.4	3.1
New methods exhilarating and challenging[a]	1.7	2.8
Worry about using the correct methods[a]	4.0	4.5
Developing good working conditions is crucial[a]	1.3	2.1
Enjoy experimenting with new things[a]	2.2	3.2
Financial and marketing skills better than parents[a]	1.8	2.2
Long-term planning better than parents[a]	1.9	2.5
Objectives different from parents[a]	2.3	3.0
Objectives different from grandparents[a]	2.4	3.2
Often challenged, requiring new decisions[a]	1.9	2.6
Managed to largely achieve goals[a]	1.5	2.0
Luck is really good management[a]	1.8	2.2
I find most employees work hard/do a good job[a]	1.8	2.3
Very determined and make things happen[a]	1.7	2.1
Management of technical issues[b]	7.7	6.8
Management of labour/contractors[b]	7.7	6.6
RELATIONSHIPS WITH OTHERS		
Enjoy manager group meetings[a]	3.1	3.5
Speak and involved in group meetings[a]	2.9	3.6

Continued

Table 7.1. Continued

	Mean value	
LEARNING ('field days' refers to visiting other similar businesses)		
Attending field days is vital[a]	2.6	3.7
Learnt a lot of technical knowledge in education[a]	2.8	3.4
Learnt a lot of technical knowledge from field days[a]	2.3	3.1
Learnt a lot of technical knowledge from reading[a]	1.8	2.1
Learnt a lot of technical knowledge from courses[a]	2.8	3.8
Learnt a lot of technical knowledge from advisors[a]	2.3	3.2
Learnt a lot of technical knowledge from reps[a]	3.0	3.6
Learnt a lot of financial knowledge from field days[a]	3.5	4.1
Learnt a lot of financial knowledge from courses[a]	3.2	4.0
Learnt a lot of financial knowledge from advisors[a]	2.4	3.0
Learnt a lot of financial knowledge from company reps[a]	3.6	4.1
OBJECTIVES		
Important to plan for leisure and holidays[a]	1.7	3.1
Minimizing pollution is important[a]	1.3	2.1
Business expansion absolutely necessary[a]	2.9	3.8
TECHNICAL FACTORS		
Years in current business	19.8	27.9
Years managing current business	18.1	21.6
Years working on business before manager	5.8	8.2
Age category with six options (1 = '< 25 years' to 6 = '> 65 years')	4.1	4.7
Output per production unit (ave.)	1005	814

[a]Scoring scale from 1 = 'true' to 5 = 'not true', meaning the lower the score, the closer to agreeing with the statement.
[b]Coded from 1 = 'poor' to 10 = 'excellent'.

can be achieved through concentrating on techniques such as relaxation and sharing the load. Learning to 'keep your cool' is also beneficial. It can also be shown that the 'high experience' managers are not that concerned at reducing risk. Also relevant is that determination and a belief that 'good luck' is made by the manager, are both valuable traits. Similarly, managers who enjoy working with new approaches, and find themselves 'challenged' at times, tend to have more useful 'experience'.

It is also important to get along with, and enjoy, other people. When it comes to formal learning, attending every opportunity to learn as well as paying close attention make contributions. All sources of knowledge are highlighted – courses, field days, written material, as well as consultants and advisors. And it is also worth noting that 'high experience' managers believe having proper holidays and leisure are part of a successful life. They are also keen on minimizing pollution as part of being a responsible citizen, but also on expanding their businesses.

In total contrast to intuition is what the theorists call the 'rational' decision-making process. Being quite the opposite, it is characterized by:

- rationality;
- intentionality;
- rules; and
- reflection (which is, however, also informally involved in intuition).

The use of intuition involves processing many pieces of information simultaneously, whereas rational decision making operates sequentially and linearly. These rational decision-making processes were outlined in Chapter 4. In reality, through simplifications, there is a continuum between rational decision making through to the use of intuition. This situation is expressed in Fig. 7.1.

The theorists would have decision makers operating rationally through using a stepped process consisting of:

- listing all the alternatives;
- gathering information about each;
- calculating the impact on the objectives of following each alternative;
- including calculating the impact of risk; and finally

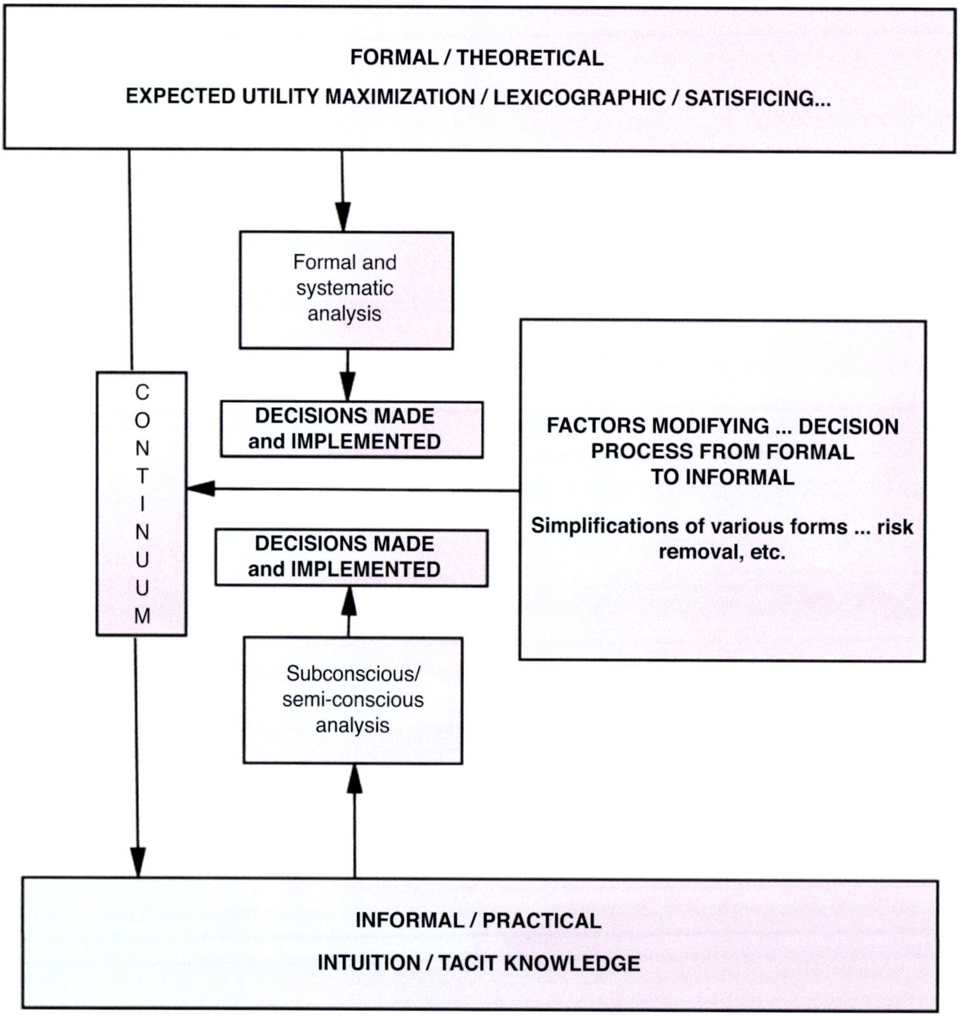

Fig. 7.1. Decision process continuum – formal to intuition. (Modified from Nuthall and Old, 2018.)

- comparing the results to select the alternative achieving the highest level of objective outcomes.

This is the process, theorist's note, of maximizing a business's 'expected utility'. Of course, this is just the first step, for the decision then has to be implemented through the 'plan, observe, replan' cycle as risk and uncertainty are played out.

While the original theorists emphasized maximum utility, others noted the reality of decision makers' objectives meant the maximization approach was infrequently followed. The concepts of satisficing and lexicographic approaches surfaced, and others too. Satisficing meant managers sought decisions providing satisfactory levels of their objectives in contrast to maximum levels. Searching for the maximums was just too complicated and time-consuming for many. In effect, they were happy with a satisfactory income, satisfactory leisure time, and so on (Chapter 4 contains a more detailed explanation of 'satisficing').

It was also suggested some decision makers put objectives into priority order (lexicographic approach), sought decisions that first achieved the requirement for their first objective (perhaps cash), then maintained this level as they sought the required level of the second most important objective (perhaps leisure), and then on to the third, while at the same time, in adjusting plans, they assured the previous attainments were still met. Eventually all resources are used up with perhaps several lowly objectives not being met at their minimum desired level.

All the elements of so-called rational decision making are represented in the top box of Fig. 7.1. But as has been noted, the reality is that most decisions are made through intuition. This implies that simplifications are brought to bear on the rational model until the decision maker develops in their psyche a process leading to what can be called 'rules of thumb' (Kelly's (1992) 'constructs') making up their intuitive set of decision systems which automatically come into play.

For example, the decision maker might automatically, through their experience, remove from consideration many of the theoretically possible alternatives as it is clear they will not meet minimum requirements. And some will not even be feasible such as, for example, a very risky alternative with a high chance of not meeting minimum 'cash for living' requirements.

Managers in particular will fall somewhere along the continuum line in Fig. 7.1 but with a strong tendency to move into intuitive decision making. However it will depend on the decision being faced at any one time. For a situation never met before, for example moving into a completely new product (perhaps grapes on a mixed business), more of a formal assessment might take place. Once experience is developed, less formality will be used as the successful rules of thumb develop.

It is also relevant in primary production and rural businesses in general, particularly due to the uniqueness of each unit's objectives, that on businesses with similar resource set-ups (e.g. soil qualities and similar climates) there are usually a number of plans and systems that will produce satisfactory outcomes. This means each manager will have their unique set of intuitive systems, none of which can necessarily be described as being superior to the others (Nuthall, 2012) but individually work well for each individual manager.

Modelling Intuition

Introduction

The research work on experience described earlier does suggest many of the variables likely to be important in intuition. To understand the development and use of intuition, it is important to isolate the variables influencing a decision maker's intuition. However, humans use intuition in many aspects of their life, not all of which are important to business management. For example, intuition involved in gourmet cooking is unlikely to be of interest compared with, say, intuition used in judging the qualities of someone newly met. The latter is often refined as more experience of the person is obtained.

In the case of primary production, the interest is primarily on the variables important in the daily decision-making life of a manager. The example above of intuition in the assessment of humans might well be important as part of the range of intuitive skills required.

As business researchers have been interested in intuition for decades, the other place to turn for important variables is the research literature. Nuthall (2018) presented a review and produced a summarizing table, a simplified version of which, taken from Nuthall and Old (2018), is shown below (Table 7.2).

The two sets of information, the research on experience and the literature, are not totally consistent. For example, the literature does not mention the decision maker's parental influences which is clearly an oversight. Nevertheless, the data available for assessing intuition as presented below does not directly encompass these parental variables so this must be borne in mind when assessing the model. However, indirect parental influences abound.

Similarly, while the main topic areas are covered in the experimental information, some of the specific questions were not available. For example, the information on learning was not complete, but in contrast additional areas were. As discussed earlier, the personality and attitude information give rise to what is referred to as the manager's 'management style'. This reflects the manager's personality as expressed in her or his management approaches (as explained in Chapter 2).

The information also allowed estimating the manager's 'locus of control' (see Chapter 3). It will be recalled this is a measure of the control over outcomes each manager believes they have. Some believe they have little due to the uncertain weather, markets and other variable factors, while others believe differently. If a manager believes minimal control abounds they will have quite a different view to making decisions in contrast to managers with a good control belief.

The literature review also emphasized the importance of decision feedback and review using a critical outlook and attitude, but this does not feature in the experience factors. This is clearly a further oversight.

Observation and anticipation skills are other variables not directly in the analysis of experience that logically you would expect to be important in intuition. Decisions must rely on an assessment of the current state of both the business and its assets, as well as the surrounding economic environment (such as the markets), using successful and accurate observation. And successful anticipation of the outcomes that might occur if decision 'A' is taken in contrast to, say, decision 'B' (and all the other

Table 7.2. A list of likely variables, grounded through the literature, important in explaining intuition. (From Nuthall and Old, 2018.)

General area[a]	Specific variables mentioned in the literature[a]
Experience, feedback and repetition	Willingness to learn; learning style; repetition of experiences; degrees of experimentation; feedback (frequency, coverage, extent, accuracy, timeliness)
Training and mentoring	Extent, form and content of training; finding and using 'masters'; extent of mentoring and its quality and form; skill at using the 'scientific method'
Reflection and self-critique	Hours spent on reflection; structured reviews; quality of reviews; use of benchmarking; extent and type of records; use of diaries and reviews; making sense of incidents; assessment of strengths, weaknesses, opportunities and threats
Intelligence and education	Practical intelligence; education – type and extent
Personality	Components of the five-factor personality model
Objectives and risk attitude	Strength and type of objectives; risk preference
Observation and anticipation skills	Observation skill level in each area of relevance; mental simulation skills in each area
General	Ability to change attitudes and systems; leadership skills; improvement process

[a]Column 1 lists the important summarizing factors and column 2 lists specific variables within each factor as suggested by the literature.

possibilities) must be a factor in useful intuitive decision making, and, indeed, any other kind of decision-making process (successful intuition involves an ability to 'mind simulate' likely outcomes). Skills in these areas will always be honed through a review of experience. These are all factors mentioned in Chapter 5.

A model of intuition incorporating the variables highlighted

Figure 7.2 incorporates the hypothesized intuition-influencing variables as outlined together with an additional influence not mentioned so far. This is labelled 'decision theory knowledge' that logically must influence successful decision making. For some reason this does not feature in the literature, nor in the experience work, which is again clearly an oversight.

By 'decision theory knowledge' is meant a knowledge of all the decision rules that have been proven to be optimal. For example, a study of business-management-oriented production economics (Nuthall, 2017) shows inputs should be applied to a technical unit (a particular moulding machine, perhaps) until the cost of the last input applied just equals the additional return achieved from this last unit. This decision rule (marginal return must equal marginal cost)

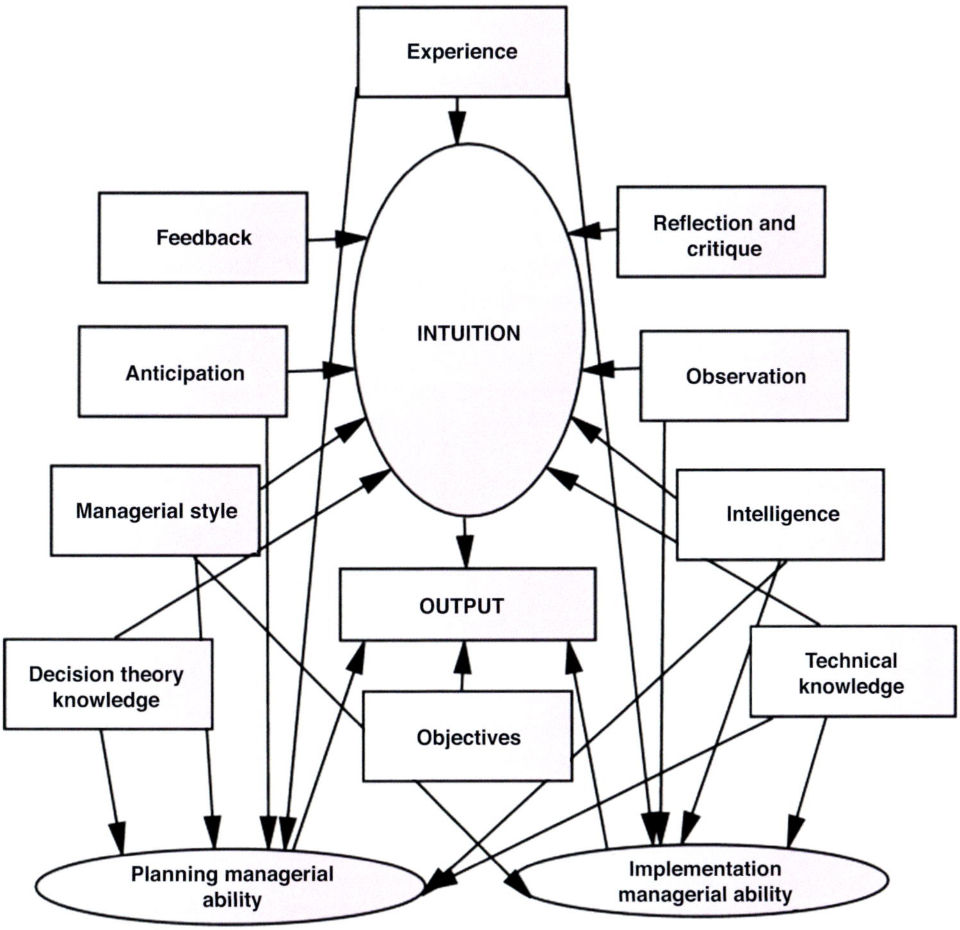

Fig. 7.2. An outline of a model explaining contributions to decision making and the resultant output emphasizing intuition, planning and implementation (basic managerial ability) as the core contributors to success. (From Nuthall and Old, 2018.)

stems from the diminishing returns nature of most production. There are many other provable decision rules that a manager should be familiar with. However, while a manager may understand many of these they probably cannot voice them in decision theory terminology (e.g. the use of the opportunity cost principle: the cost of a resource is equal to the return achievable in the next best use). Many books cover the ideas (e.g. in story-based form, see Nuthall, 2016).

The centre of the hypothesized model is the business's 'output', which in this case, as shown below, was a range of factors including net profit per labour unit (a measure comparable across different business types and resource qualities), net assets per person, average yearly change in profit over 5 years (to even out weather and market variations) and the average yearly change in net assets over 5 years (which reflects outcomes over and above declared profit). Also relevant should be outputs such as leisure but these were not recorded other than through measurements of the manager's objectives. Financial outcomes, relative to manager group levels, should reflect these issues after allowing for the manager's productive efficiency, which was measured as a relative factor.

The model (Fig. 7.2) features both rectangles and ellipses. The variables shown in the rectangles are directly observable whereas the others in ellipses are surmised from the rest of the data and are calculated using the assumptions portrayed by the model. Thus, intuition, which is emphasized through its size and position in the diagram, is not directly observable but is inferred from the observed variables. The model, of course, does depend on the hypothesis that intuition is dependent on the variables listed which, given the experience data, is a reasonable assumption.

It is also hypothesized that outcomes are dependent on a manager's planning ability as well as their ability to implement the plans. In both cases, it is reasonable to construe that intuition is important in these areas. Neither variable (planning abilities, implementation abilities) is directly observable and consequently cannot be directly measured.

The model presented in Fig. 7.2 has been simplified for ease of exposition. In many cases the rectangle-based variables are conglomerates of several sub-variables. For example, the manager's objectives are made up of several factors such as risk attitude, leisure requirements, and so on. Similarly, technical knowledge, which is clearly critical to success, covers skills in many sub-areas. The same applies to most rectangles, as will be clear in the definitions listed below.

The other important aspect to Fig. 7.2 is the linking arrows. Each represents the hypothesized direction of influence and, in a calculational sense, represents an equation specifying the details and strength of the influence. The critical information is the variables' strength data, as this then implies the variables that should be featured in improving intuition.

The model as portrayed suggests intuition development is a static operation. In reality, intuition, and the influence of all the variables, will be a dynamic system exhibiting continual change. However, the data available to analyse the model is a snapshot of managers' situations at the time they provided the information. To assess the dynamic impacts of intuition it would be necessary to obtain information every few months and use the sets as a series of model snapshots. These could then be analysed to see the dynamic changes and whether the influence of the variables had any untoward time-dependent impacts. This must await further research.

Analysis methods

Given a snapshot set of data the obvious way to assess the importance of each variable is the simultaneous calculation system known as structural equation modelling (SEM) (Bowen and Guo, 2012). This uses maximum likelihood methods to estimate the linear regressions associated with each arrow and leads to understanding the importance of each observed variable in their simultaneous impact on each 'latent' (listed in the ellipses) variable. The solution is found using an iterative process.

In a subsequent section all this information is presented, leading to important conclusions. As the analysis also showed that productive efficiency was very important, as might well be imagined, a separate regression equation was developed to highlight the important factors in achieving high efficiency in output production.

This information is also presented as it is clearly important to intuitive skill.

To reinforce the conclusions on all the important variables influencing intuition, the results of the SEM analysis were used to group the managers into three clusters. These included the managers with successful intuition, those with poor intuition and all those 'in between'. Then, to highlight the differences between the 'good' and the 'bad', a comparison was made of the 'successful intuition' against the 'poor intuition' managers showing the percentage difference in the variables determining intuitive skill.

It will have been noticed the output measures were mainly financial connotations. Yet it has been stressed that managers seldom have purely financial interests. To allow for this, the managers' objectives feature as an input variable which then mediates the assessment of the financial criteria.

Details of the Variables and Data Collection

As noted, each core variable from the model (Fig. 7.2) is made up of a number of sub-variables, each of which is listed below. The data for all these variables was obtained from a mail survey of 2268 randomly selected businesses stratified by type, area and region to ensure a good sample. The response rate was 36%, giving 818 successfully completed questionnaires. Details of the questionnaire are given in Nuthall and Old (2014), which is accessible online.

Assessing each sub-variable required answers from the managers to assess the many facets of each; this resulted in a questionnaire that was eight pages long. For example, the questions to assess a manager's 'management style' involved 25 statements. An example (taken from Chapter 2) is 'For most things you seek the views of many people before making changes to your operations'. The managers were asked to rate each one's truthfulness on a five-point scale (1 = 'true' through to 5 = 'not true'). The full list of questions is given in Appendix 2A in Chapter 2. Similarly, a set of 20 questions was used to assess a manager's goals and objectives (such as 'It is very important to ensure employees enjoy their jobs') as listed in Chapter 3. For the Locus of Control, the full set of questions (e.g. 'I never try anything that might not work'), of which there were 19, are also given in Chapter 3.

It will be recalled that Chapter 2 contains a discussion on personality emphasizing the five-factor model as summarized by the word OCEAN, the letters of which stand for one of the five factors (Matthews and Deary, 1998). OCEAN gives (O)penness, (C)onscientiousness, (E)xtraversion, (A)greeableness and (N)euroticism (which is better called 'anxiety'). For managers, the research on these factors, as explained in Chapter 2, showed six factors were discovered for defining a manager's management style.

Notwithstanding this work, it should also be noted there are other theories on personality, such as that proposed by Myers and Briggs (Keirsey, 1998), but generally the five-factor model tends to dominate.

To ensure strong measures of each construct, a large number of questions were needed. For parsimony, it was important to summarize them down to the basic ideas expressed, through allowing for the correlations between the detailed questions. This process is known as factorization, an explanation of which can be found in most basic statistics texts. In this reduction process, choices exist over how far down to the basic construct to proceed. In this case factors had what are called eigenvalues of 1 or greater and are presented after a Varimax rotation which gives more obvious logic to the figures.

These analyses produced the following list of variables (taken from Nuthall and Old, 2018) for each construct shown in Fig. 7.2. Each sub-variable was given a name based on the main questions making up each 'factor', as they are called.

- **Intelligence**: level of formal education achieved, grades obtained and also gender and age.
- **Management style**: Style 1 – 'consultative logician community', Style 2 – 'correctness seeker', Style 3 – 'consultative logician family and friends', Style 4 – 'conscientious planner', Style 5 – 'thoughtful creator', Style 6 – 'benign manager'.

- **Experience**: Exp. 1 – 'years of experience', Exp. 2 – 'people centric experience', Exp. 3 – 'decision allocator', Exp. 4 – 'major problem experience', Exp. 5 – 'mistake learner'.
- **Objectives and risk**: Obj. 1 – 'balanced', Obj. 2 – 'risk remover', Obj. 3 – 'way of life', Obj. 4 – 'reluctant manager', Obj. 5 – 'community supporter', Obj. 6 – 'family supporter'.
- **Anticipation and simulation**: Antic. 1 – 'plan and act', Antic. 2 – 'imperative planner'.
- **Feedback and mentoring**: Feedback 1 – 'formal 'groupy', Feedback 2 – 'informal conferrer', Feedback 3 – 'family/friend discusser', Feedback 4 – 'professional conferrer seeker'.
- **Observation skills**: Obs. 1 – 'all-round observer', Obs. 2 – 'diligent record keeper', Obs. 3 – 'system observer', Obs. 4 – 'administration observer'.
- **Reflection and self-critiquing**: Reflect 1 – 'strict reflector', Reflect 2 – 'worrisome reflector', Reflect 3 – 'careful reviewer'.
- **Technical and system knowledge**: Tech. 1 – 'knowledge seeker', Tech. 2 – 'output achiever', Tech. 3 – 'determined technocrat'.
- **Decision theory and analysis knowledge**: Dec. 1 – 'theory compliant', Dec. 2 – 'theorist practitioner'.
- **Productive efficiency**: physical output relative to peers.
- **Outputs**: average net profit per person, average net assets per person, average yearly change in net profit over 5 years, average yearly change in net asset levels over 5 years. The profit and net asset figures were divided by the number of people working in each business to provide comparability.

The actual model using all these variables ended up being slightly different from Fig. 7.2 (which was created after the calculations). In the interests of parsimony and ease of calculation, the model used only had one ellipse for 'managerial ability', but it did have 'productive efficiency' as a variable representing 'implementation ability' as an observed variable in contrast to latency. One or two other variables were also introduced such as an index of business type to allow for the different environmental issues.

Analysis of Data and Results

Several sets of analyses, as noted above, are presented. The first is one of the maximum likelihood solutions to the intuition model; the next is a comparison of the managers' 'personological', and other, variables for 'high intuition' managers compared with the less skilled managers; and the final set of information presents the parameters of the regression explaining productive efficiency.

The SEM model

Figure 7.2 model represents lines of influence, each one of which is assumed to be a linear regression. The coefficient associated with each line is presented in Table 7.3 after it was standardized to enable direct comparisons. This process removes the impact of the different units each is measured with (the raw coefficient is divided by its standard deviation). On the left of the table is each pair of variables encompassing each line – the originating variable and the variable on which it impacts. The last column gives the significance probability of each standardized value. Thus, for example, the first line gives the impact of manager's gender on their educational grades which is highly significant as shown by the 0.008 probability (0.8%). As another example, the fourth line shows managerial style as a grouping has a relative 1.743 impact on managerial ability with a significance probability of 6.3%. In interpreting the significance probability it is worth taking note of any influences less than 0.200 for this means there is an 80% chance of correctness.

In assessing the data some unusual negative coefficients will be noted. However, on considering how the variables were measured most become logical. For example, the management style variables are based on one reflecting 'truth' relative to five for 'non-truth', meaning that lower values are often desirable. Similarly, managerial ability was measured using the same

Table 7.3. Parameters for the intuition Structural Equation Model based on the model in Fig. 7.2. (From Nuthall and Old, 2018). Variables in UPPER CASE are a grouping covering one of the rectangles in Fig. 7.2. This approach reduces the table size and provides the critical information.

Variable pair (influence left to right)	Standardized coefficient (actual or sum if a grouping of variables)	Significance probability (ave. if a group)
Gender to grade	0.093	0.008
Gender to education	0.217	0.000
Age to education	–0.176	0.000
STYLE to managerial ability	1.743	0.063
Education to managerial ability	–0.259	0.000
Grades to managerial ability	–0.191	0.000
Managerial ability to DECISION THEORY	1.086	0.000
Managerial ability to TECHNICAL KNOWLEDGE	0.863	0.164
Managerial ability to ANTICIPATION	0.381	0.068
Managerial ability to OBSERVATION	0.960	0.130
EXPERIENCE to intuition	0.019	0.434
REFLECTION to intuition	0.163	0.232
OBSERVATION to intuition	–0.035	0.300
Grade to intuition	0.114	0.084
DECISION THEORY to intuition	0.394	0.079
ANTICIPATION to intuition	0.128	0.185
FEEDBACK to intuition	0.015	0.534
TECHNICAL KNOWLEDGE to intuition	0.945	0.076
OBJECTIVES to profit change percentage	0.148	0.554
OBJECTIVES to asset change percentage	0.142	0.177
OBJECTIVES to assets per person	0.403	0.051
OBJECTIVES to profit per person	0.273	0.053
Resource quality to profit change	0.007	0.846
Resource quality to asset change	–0.026	0.455
Resource quality to productivity	0.016	0.452
Resource quality to assets per person	–0.039	0.195
Resource quality to profit per person	–0.041	0.186
Managerial ability to profit change	–0.169	0.000
Managerial ability to asset change	–0.243	0.000
Managerial ability to productivity	–0.469	0.000
Managerial ability to assets per person	–0.512	0.000
Managerial ability to profit per person	–0.507	0.000
Intuition to profit change	0.035	0.366
Intuition to asset change	0.175	0.086
Intuition to productivity	0.919	0.071

Continued

Table 7.3. Continued

Variable pair (influence left to right)	Standardized coefficient (actual or sum if a grouping of variables)	Significance probability (ave. if a group)
Intuition to assets per person	0.198	0.081
Intuition to profit per person	0.296	0.075

scale, which is why the managerial impacts on outputs are negative.

There are a number of standout influences expressed in these results. The largest standardized coefficient is the impact of managerial style on managerial ability, notwithstanding this is a gathering of the five individual variables. This clearly reflects the influence of the human factor on ability as is stressed throughout this book.

Given this chapter is highlighting intuition, the main variables of interest are those hypothesized as impacting on intuition. It will be noted if the coefficients are shifted so the lowest is zero, the sum of the intuition variables is 3.473 relative to a sum of 0.66 for 'managerial ability', which is set up to reflect planning ability per se. This clearly reflects the importance of intuition.

Also of interest is the impact of the objectives on the financial outputs. These, all being less than 1, reflect the ameliorating impact they have on the straight profit motive with the biggest impactor being on asset levels. Also relevant to note are the technical knowledge influences, emphasizing how important technology is (as would be expected).

In terms of a manager improving her/his intuitive skills, the key knowledge is given by the contributing variable coefficients. In order of importance they are:

- technical knowledge (0.945);
- decision theory knowledge (0.394);
- decision reflection and critique (0.163);
- anticipation skills (0.128);
- experience (0.019);
- feedback (0.015); and
- observation skills (–0.035).

In assessing these values their importance must be tempered by their significance probabilities. The feedback coefficient is not as high as might be expected given its low significance – you might in reality expect it to be higher.

Logic would suggest both technical and decision theory knowledge to be critical and essential to good decision making and this is clearly borne out by the results. The remaining factors are of a lesser importance but nevertheless still critical components of intuition, with reflection and critiquing being the leading factors. Experience per se follows but given all the other variables related to experience this is not surprising because experience by itself is not valuable, it is how it is used that is important. Relative to the model in Chapter 3, this model has many more variables theorized to be part of intuition and consequently experience by itself has relatively little impact.

Observation and anticipation are also very important to managerial ability in its planning capacity. This makes total sense, and, of course, as noted, technical knowledge is important through its impact on planning ability. And on top of these relationships is the critical nature of intuition to technical efficiency (productive efficiency). More information on the basic factors influencing productive efficiency is presented below relative to the regression exploring this variable.

A comparison of managers with high as against low intuition abilities

To obtain a better idea of the relationship of the individual sub-variables to intuition, the results from the SEM were used to develop an equation to predict the unobserved 'intuition' for each manager. This enabled separating out the three intuition skill groups which, of course, reflect the distribution of intuition, as shown in Fig. 7.3. It is close to normality and confirms, with respect

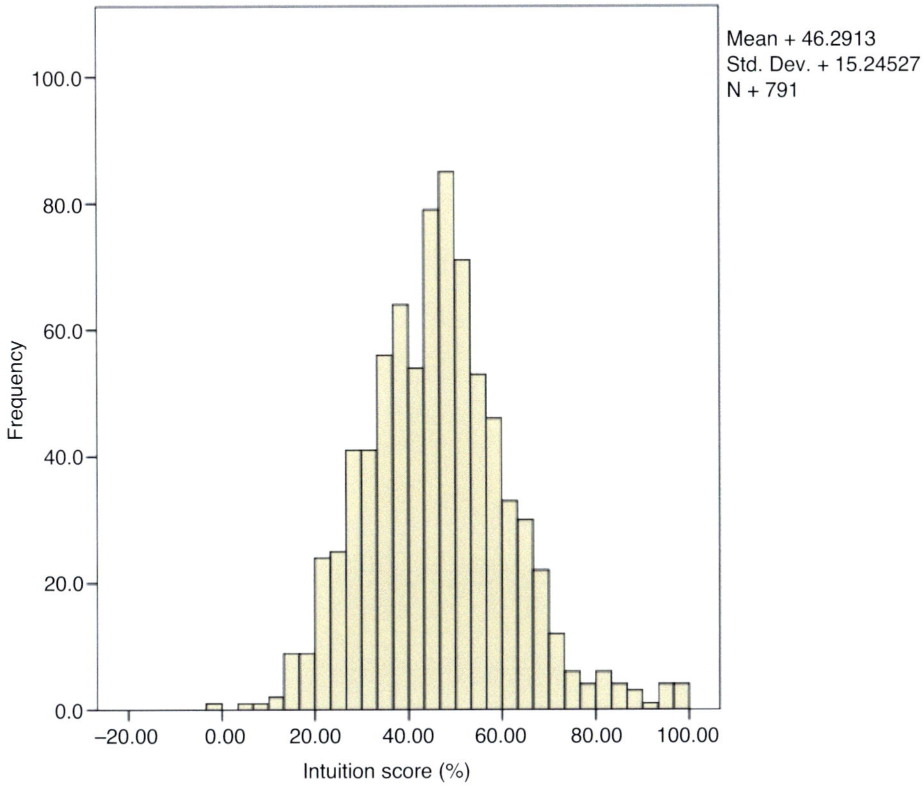

Fig. 7.3. Frequency distribution of managers' intuition score. (From Nuthall and Old, 2018.)

to the statistical tests, that the constituent variables approach normality.

The group information presented in Table 7.4 are the average parameters of managers with an intuition score of greater than 60%, and less than 41%. The percentage difference between their parameters is shown, as is the significance probability of the differences. The figures to take note of will have a difference of at least 100% and a significance probability of less than 0.200. A large number of the differences between the groups are quite significant. Indeed, the significance levels are better than those shown for the individual variables in the SEM results. Thus, the comparison data more than reinforces the SEM information.

Of the managerial style factors, they all have large percentage differences following the book's general hypothesis that the human factor matters. However, the significance of the differences for the 'correctness seeker' and 'thoughtful creator' is poor. While six style variables are used, their relationship to the five-factor personality model will be clear. Overall, the results provide a manager interested in improving her or his intuition with clear personality targets to consider. For example, paying attention to ensuring conscientiousness is one clear objective.

While managers may consider their objectives fixed, it is also interesting that certain objectives influence intuition. The 'way of life' aspect to business influences intuition quite significantly.

Looking further down the list 'people critic experience' is clearly an important factor as are many other factors with over 200% differences. All the technical knowledge sub-variables are important as was expected, and similarly decision theory knowledge. The 'mistake learner' falls into the same category and is clearly related to experience in which the 'years of experience' is important.

Table 7.4. A comparison of important variables of managers with an intuition score of less than 41%, and greater than 60%. (From Nuthall and Old, 2018.)

Variable group	Variable	Percentage difference	Significance probability	Desirability (NS = non-significant)
General	Age	5	0.049	Historical
	Education level	3	0.567	Historical
	Grades	4	0.324	Historical
Latent	Managerial ability	40	0.000	Greater
Style factors	Consultative logician community	221	0.000	Greater
	Correctness seeker	932	0.682	NS
	Consultative logician family and friends	215	0.000	Greater
	Conscientious planner	315	0.046	Greater
	Thoughtful creator	675	0.464	NS
	Benign manager	220	0.000	Greater
Objective factors	Balanced	142	0.167	Greater
	Risk remover	87	0.766	NS
	Way of life	398	0.000	Greater
	Reluctant manager	272	0.000	Lesser
	Community supporter	437	0.002	Greater
	Family supporter	107	0.802	NS
Anticipation factors	Plan and act	186	0.431	NS
	Imperative planner	236	0.000	Greater
Feedback factors	Formal 'groupy'	180	0.121	Greater
	Informal conferrer	236	0.358	NS
	Family/friend discusser	220	0.039	Greater
	Professional conferrer seeker	258	0.000	Greater
Observation factors	All-round observer	227	0.005	Greater
	Diligent record keeper	208	0.052	Greater
	System observer	114	0.399	NS (lesser)
	Administration observer	618	0.233	Greater
Reflection and critique	Strict reflector	271	0.320	Greater
	Worrisome reflector	229	0.000	Lesser
	Careful reflector	235	0.081	Greater

Continued

Table 7.4. Continued

Variable group	Variable	Percentage difference	Significance probability	Desirability (NS = non-significant)
Technical knowledge	Knowledge seeker	280	0.000	Greater
	Output achiever	236	0.000	Greater
	Determined technocrat	239	0.000	Lesser
Decision knowledge	Theory compliant	234	0.000	Greater
	Theorist practitioner	213	0.000	Greater
Experience	Years of experience	388	0.177	Lesser
	People critic experience	315	0.009	Greater
	Decision allocator	173	0.465	NS
	Major problem experience	150	0.406	NS
	Mistake learner	219	0.000	Lesser
Outputs	Productive efficiency	251	0.000	Top intuit >
	Percentage change profit per year	24	0.005	Top intuit <
	Percentage change assets per year	31	0.210	Top intuit >
	Profit per person	13	0.222	Top intuit >
	Net assets per person	22	0.050	Top intuit <

'Top intuit' refers to the managers in the top intuition group. Thus, for example, 'Top intuit >' means the managers in this group had a greater level of the factor in the variable referred to.

Productive efficiency

The 'productive efficiency' of the two intuition level groups is clearly very different. This variable was calculated using the different physical output levels for each business type to find the top output business. These top businesses then provided the yardstick against which all the other managers were ranked.

As has been stressed, due to the importance of productive efficiency, a linear regression was calculated to explore the important relationships with variables that might be under manager control (this is where the 'locus of control' could be important). The standardized coefficients for the regression are given in Table 7.5. It will be noted just over 70% (R^2) of the variance in productive efficiency is explained by the equation and the equation is highly significant.

Many of the variables contribute in small ways to efficiency with few being outstanding. However, one important contributor is the technical component 'output achiever' which had as its constituents variables such as field day attendance, allowing employees and contractors to 'do it their way', and a knowledge of governance systems.

In contrast, having a technical factor 'determined technocrat' as a feature was counterproductive – perhaps these people could not see the wood for the trees. Similarly, being a 'thoughtful creator' was a drawback possibly due to too much thought and experimentation and insufficient 'doing'. However, the 'knowledge seeker' achieved good results as you would hope.

Table 7.5. Standardized coefficients from a linear regression explaining the productive efficiency (R^2 = 0.723, significance probability = 0.000).

Variable	Standardized coefficient	Significance probability
Planning factor 1 – Help seeker	–0.093	0.146
Planning factor 2 – Reviewer	0.053	0.171
Objectives 1 – Balanced mix	0.010	0.769
Objectives 2 – Risk remover	0.011	0.661
Objectives 3 – Way of life	–0.067	0.028
Objectives 4 – Reluctant manager	–0.100	0.001
Objectives 5 – Community supporter	0.019	0.413
Objectives 6 – Family supporter	0.020	0.374
Style 1 – Consultative logician community	–0.046	0.230
Style 2 – Correctness seeker	–0.069	0.032
Style 3 – Consultative logician family and friends	0.006	0.878
Style 4 – Conscientious planner	–0.022	0.662
Style 5 – Thoughtful creator	–0.0463	0.000
Style 6 – Benign manager	–0.111	0.001
Feedback 1 – Formal ‘groupy’	0.082	0.187
Feedback 2 – Informal conferrer	–0.008	0.749
Feedback 3 – Family/friend discusser	0.017	0.520
Feedback 4 – Professional conferrer seeker	0.010	0.640
Anticipation 1 – Plan and act	–0.015	0.468
Anticipation 2 – Imperative planner	0.039	0.148
Experience 1 – Years of experience	–0.025	0.243
Experience 2 – People centric experience	–0.036	0.245
Experience 3 – Decision allocator	0.001	0.976
Experience 4 – Major problem experience	0.045	0.051
Experience 5 – Mistake learner	0.001	0.967
Observation 1 – All-round observer	–0.031	0.455
Observation 2 – Diligent record keeper	–0.090	0.021
Observation 3 – System observer	0.153	0.000
Observation 4 – Administration observer	0.028	0.244
Reflect 1 – Strict reflector	–0.027	0.329
Reflect 2 – Worrisome reflector	–0.071	0.109
Reflect 3 – Careful reflector	–0.060	0.072
Technical 1 – Knowledge seeker	0.203	0.000
Technical 2 – Output achiever	0.823	0.000
Technical 3 – Determined technocrat	–0.406	0.000
Decision 1 – Theory compliant	0.135	0.000
Decision 2 – Theorist practitioner	0.050	0.134

Also important is being a 'system observer', which implies standing back at times and ensuring constant inspections of the current 'state of affairs' on both the business and beyond the business gate. On the other hand, having a style encompassing being a 'benign manager' is a deterrent to efficiency.

Overall, however, it is clear attention to many attributes and factors is important and this all adds up to productive efficiency. This is a clear message and not at all unexpected.

More on Modelling Intuition – the Core Variables in Intuition

The intuition modelling so far discussed, and the associated data analysis, were all based on the results of a manager survey especially designed to explore the issue. If another set of data from another jurisdiction or set of businesses was conducted, there is, of course, no guarantee that the results would be similar. For sure the specific parameters of the variables associated with intuition would turn out to be different in their detail if not in the relative values, which is, of course, the more important issue. While the latter is most likely, it will be necessary to wait until a range of additional studies are conducted over the years to be certain.

Another factor that has been commented on is the criteria used to judge the managers' successful use of intuition. In the results reported so far the main criterion was the financial return. As is well known, rural managers and their families have a much wider range of objectives than just financial outcomes. Indeed, in many cases, if it was only the financial outcomes that mattered many rural managers would be advised to sell up and invest the money in other, mainly urban, businesses and investments. In general, the latter produce a rate of return greater than what rural businesses are producing in many cases – long-term average rates on many rural businesses hover around 4%, making it clear these other objectives provided by many rural-based businesses are important.

To overcome some of these issues, further research was undertaken using a much larger database obtained by putting together the results from three other studies that contained information for each business on the manager's characteristics so enabling combining the overall data. The additional data obtained to assess different issues in each survey covered a range of different topics that were not common to each survey and did not relate to managerial ability and intuition. This data was removed to simplify combining the relevant data into one common database. The surveys were conducted in different years and were adjusted to account for year-influenced differences.

This data set of some 1950 managers was then used to investigate intuition factors from this much wider group of rural businesses. The particular question posed was to determine the nature of what can be called the *core variables giving rise to intuitive decision ability*. It is known most managers make most decisions through the use of their intuition. As has been stressed, this is not simple everyday intuition but a decision approach which has as its base a lifetime of education, training and experience. The full details of the study are reported by Nuthall (2022) in an easily accessible research publication (use https://doi.org/10.3390/bs12110409 in a browser to obtain a full copy). The discussion which follows summarizes that study.

The report refers to this 'well honed' intuition as being 'informed intuition', or more specifically as 'objectively informed intuition' (OII for short). A review of decision-making literature summarized in the article concludes the core variables giving rise to a business person's OII are the manager's training, experience and what is called the decision maker's phenotype. Phenotype is the measure of a manager's personal business-related (BR) characteristics which is obtained from each manager in the surveys by quantifying their BR personality, education, observation and anticipation skills through obtaining their strength of belief in a wide range of statements covering all these attributes.

In a general sense, phenotype, being what you observe in a decision maker, stems from a combination of the manager's genetic make-up and the experiences encountered from a young age with its parental influences, and beyond. That is, phenotype is a combination of environmental experiences and the genetic inheritance.

In a nutshell, the hypothesis is encapsulated by the model shown in Fig. 7.4, proposed by Nuthall (2022), which is largely self-explanatory.

An additional variable not so far introduced is the 'Formal skills in planning'. This was added as a manager can have any number of OII-based plans and decisions, but if they are not implemented successfully and managed appropriately the outcomes will not do justice to the OII decisions. Nuthall (2022, p. 11) expresses this model as follows.

A general equation portraying the OII model is given by:

$$oii_t = oii_{t-1} + \Delta f\left(p_t, t_t, e_t\right)$$

where

oii_t decision maker's 'objectively informed intuition' skill level at time t (it will be noted that Fig. 7.4 has OII reflecting both planning and plan application skills)

p_t individual's business-related phenotype status, at time t, accruing from all previous periods

t_t individual's past formal training status at time t accruing from earlier time periods

e_t sum of information learnt, at time t, from the individual's past experiences

Δ change, i.e. the change in p, t and e over the last period. That is, values at time t minus values at time $t - 1$.

For the moment the units of measure are not identified.

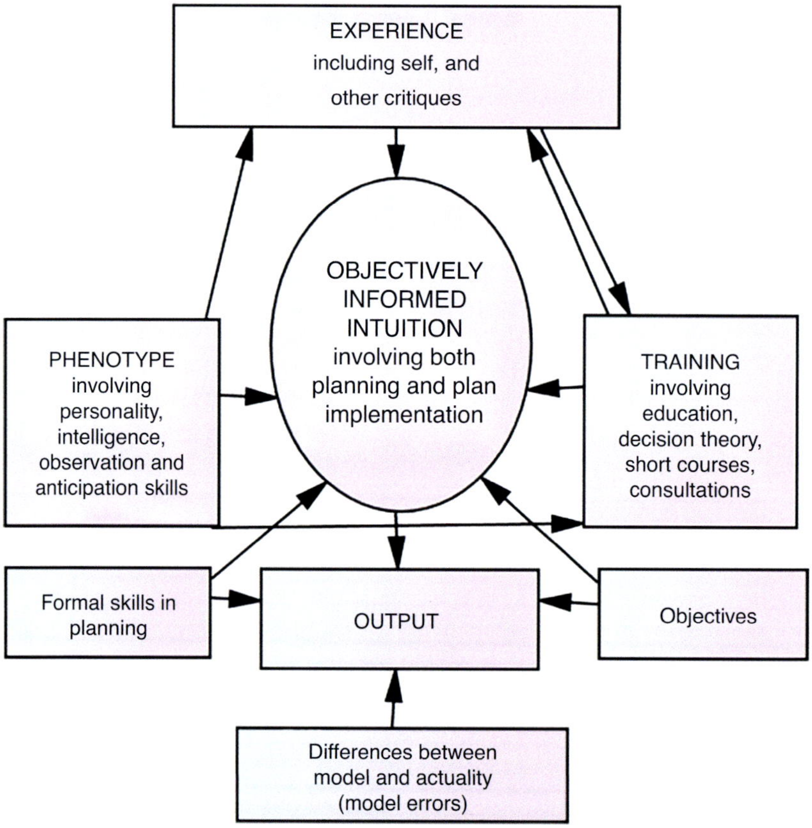

Fig. 7.4. An outline of a model explaining the contributors to 'objectively informed intuition' (OII) decision making and the resultant output. (From Nuthall, 2022, p. 10.)

To actually allow the model to be quantified a relationship between output and its precursors is required. Thus:

$$O\left(utcome\right)_{j,t} = f\left(oii_t, n_t, s_t\right)$$

where the new variables introduced are

n formal planning input

s_t situation of all environmental factors influencing success including prices, costs, resource levels

$O\,(utcome)_{j,t}$ variable reflecting the output of each distinct objective j at time t.

In the case being proposed it is assumed all decisions are made through 'objectively informed intuition' with the assumption that formal documented planning is not directly occurring (which is the main situation), and the situational variable is set at 'average' given the range of years available in the data. Thus, n_t and s_t drop from the equation. In this formulation it is assumed *oii* embodies both decision planning and application skills.

The parameter values for each variable in the analysis of the data do depend on the criteria used in assessing the quality of each manager's OII. As it turned out, the most useful criteria were the surplus cash obtained by each manager but weighted by the other factors entering a manager's objective set. And on a different tack, as physical efficiency in production is critical, something that is important – no matter the other objectives – is productivity.

The Structural Equation Model used provided the result, when averaged across these two important objective criteria, that phenotype contributed 50.7% to output and training 39.1% with experience at 10.2%. This shows the importance of the genetic heritage obtained from the earlier generation as well as the early environment, all factors giving rise to a manager's phenotype. It also shows experience has somewhat less of an impact than might be expected. However, experience per se is of limited use unless properly analysed to learn the lessons held in the material – this is where phenotype is so important.

In contrast, when the earlier research on intuition is compared with these results, while overall the results are similar, the experience factor has a higher contribution. Using the data from the comparison of the various characteristics used in the first set of experiments and combining the percentage difference figures that were significantly different between the variables from the managers with high and low intuition level that lead to the phenotype, training and experience, it was found phenotype contributed 40.3% to successful intuitive decision making, training 29.5% and experience 30.2%. While this shows phenotype as again being the most important variable, it also shows training and experience have similar contributions. This higher figure for experience might well be more in line with reality. However, it won't be until many more studies are conducted that the true figures for each will firm up. On the other hand, it would not be expected to have the figures all relatively identical for all types of business and in all areas. And it must also be recognized that in judging intuitive decision efficiency the first experiments used financial returns as the criteria whereas additional objectives were allowed for in the combined data analysis.

However, when considering the efficiency of physical output, the relative contribution to high levels of efficiency shows much more similarity in both sets of experiments. It must be recognized that physical efficiency will always be a precursor to monetary efficiency. Maximum output for given inputs must always be sought.

Improving Managers' Intuitive Skills

The results of the analyses make it clear the variable constructs hypothesized to be important in the development and maintenance of a manager's intuition have all turned out to provide contributions. If the standardized coefficients of each construct in the first study are summed, allowing each to be expressed as a percentage contribution, besides the contribution of decision theory and technology knowledge, the contributions are:

- feedback 4.2%;
- experience per se 5.3%;
- anticipation 35.5%;
- observation 9.7%; and
- reflection 45.3%.

This information makes it very clear where any manager should give her or his attention. Of course, this is only one study, and others may well discover different percentages, and no doubt they will as each set of manager data will be unique.

In addition to the SEM analysis, for comparative purposes, a regression was used to explain intuition. This led to the relative importance of the core variables as feedback at 14.3%, experience at 19.8%, anticipation 5.5%, observation 29.3%, and reflection at 31.1%. Other than anticipation, the order is similar. However, a linear regression does not reflect the interaction between the variables that SEM allows. In the case of the SEM model, decision theory and technical knowledge combined led to more than a 70% contribution relative to the others, whereas for the regression there was a 25% contribution.

While these are quite different contributions, it is clear these knowledge areas are important. The variation in the results emphasizes just how important it is to use what is a logical model – which, in this case, must be the SEM-based construction with its simultaneous impact of the different components of intuition.

In the Nuthall (2022) study, the second piece of research on the core variables giving rise to OII excellence, the relative importance of training and experience was determined. However, detailed aspects of these two significant factors were not explored. To provide more detail further study was conducted by factorizing the variables making up each core variable based on constituents with eigenvalues greater than one with a Varimax rotation. Where a decision maker wishes to work on improving their level of OII efficiency, a knowledge of the importance of each factor gives direction to working on improving the important factors associated with training and experience. This second study, it must be remembered, used efficiency criteria allowing for all a decision maker's objectives, not just the cash returns.

In the *training variable* case it was found three important factors could be isolated. Based on their constituents these were named:

- 'developer of formal planning skills';
- 'gatherer of benchmarking skills'; and
- 'studier of management principles'.

For the *experience variable* factors seven important aspects explaining 52.2% of the variance were isolated and named:

- 'seek help from others';
- 'considerate employer';
- 'perseverance skills learnt';
- 'lessons from bad luck';
- 'extensive and careful planner';
- 'discuss and mull over lessons from experience'; and
- 'care and critiquing of others' suggestions'.

The training factors explained 41.1% of the variance.

Training Factors' Impact on OII Ability

Tables 7.6–7.9 allow assessing the importance of each of the training factors. All the respondents were divided into four groups based on their level of OII skill, where the criteria were based on a range of attributes. Each table lists the attribute used in dividing the respondents into groups and then correspondingly presents the average value for the top and bottom groups allowing, by comparison, determining the desirable value of each attribute. These form benchmarks to work towards in any training programme.

Each of the training factors defines general skills believed to be required for OII excellence by the respondents. A wide range of planning skills features highly in the managers' minds. To interpret the data it is important to assess the scores relative to the questions that make up the three factors. This reveals the following interpretations:

- Developer of formal planning skills – if the factor score is relatively low, this means the manager highlights putting effort into learning formal planning skills.
- Gatherer of benchmarking skills – work on making sure the business systems compare well relative to other managers is reflected through a low score.
- Studier of management principles – a low score indicates the manager values training that leads to understanding the principles.

It will be noted this last training factor does not have significant score differences between

Table 7.6. Cash change percentage OII scores. Average factor scores for training factors relative to the top and bottom groupings of OII ratings based on cash change percentage OII scores.

Training factor	Top 25%	Bottom 25%	Difference	Significance	Range
Developer of formal planning skills	0.322	–0.287	0.609	<0.001	–2.842–3.539
Gatherer of benchmarking skills	0.544	–0.501	1.045	<0.001	–3.282–6.396
Studier of management principles	0.092	–0.202	0.294	<0.001	–3.086–4.081

Table 7.7. Productivity OII scores. Average factor scores for training factors relative to the top and bottom groupings of OII ratings based on productivity OII scores.

Training factor	Top 25%	Bottom 25%	Difference	Significance	Range
Developer of formal planning skills	0.367	–0.320	0.687	<0.001	–2.842–3.539
Gatherer of benchmarking skills	0.574	–0.536	1.110	<0.001	–3.282–6.396
Studier of management principles	0.086	–0.177	0.263	<0.001	–3.086–4.081

Table 7.8. Objective-weighted cash change. Average factor scores for training factors relative to the top and bottom groupings of OII ratings based on cash change by objective-weighted scores.

Training factor	Top 25%	Bottom 25%	Difference	Significance	Range
Developer of formal planning skills	0.221	–0.202	0.423	<0.001	–2.842–3.539
Gatherer of benchmarking skills	0.274	–0.302	0.576	<0.001	–3.282–6.396
Studier of management principles	–0.084	–0.092	0.176	<0.453	–3.086 – 4.081

Table 7.9. Objective-weighted productivity. Average factor scores for training factors relative to the top and bottom groupings of OII ratings based on productivity by objective-weighted scores.

Training factor	Top 25%	Bottom 25%	Difference	Significance	Range
Developer of formal planning skills	0.291	–0.290	0.581	<0.001	–2.842–3.539
Gatherer of benchmarking skills	0.374	–0.065	0.439	<0.001	–3.282–6.396
Studier of management principles	–0.053	–0.065	0.118	<0.424	–3.086–4.081

the top and bottom managers other than for the OII ratings based on the cash outcome. Furthermore, the score differences are minimal relative to the possible score values (see ranges). In effect, this factor is not a training issue of great importance.

Surprisingly, for the other two training areas the less successful managers tend to have a lower rating than the more successful managers. Either the questions used were misleading in some way, or indeed these main training factors are not that relevant. However, it needs to be noted that while the differences are all highly significant, relative to the ranges possible for each, the differences between the two groups are not large.

Overall, the lessons here are not remarkable other than to note the respondents' answers do not reflect clear differences for the training aspects covered. However, other factors may well be important. Most managers have training providing at least a modicum of all these skills. Common logic would lead to ensuring a manager would have reasonable abilities in each of these areas.

It is also relevant that a manager does not necessarily need to upskill her or his planning skills, nor their understanding of management principles, if they simply work towards achieving benchmarks obtained from the better managers as their guide to good management. It is not known how prevalent this approach is, other than to comment it is certainly not hard to find examples of this 'follow the leader' approach. Such an approach may lead to poor results for any particular business if in reality the businesses are not directly comparable. This situation is not uncommon – most businesses are unique, particularly over objectives.

Experience Factors' Impact on OII Ability

Tables 7.10–7.13 allow assessing the importance of each of the experience factors. Again, all the respondents were divided into four groups based on their level of OII skill where the criteria were based on a range of attributes. Each table gives the attribute used in dividing the respondents into groups and presents the average value for the top and bottom groups allowing comparisons of the desirable value of each attribute.

The greater the experience factor score the greater the average output, reflecting that experience matters in achieving high-level outcomes in all the success measures used.

The direction of the questions leads to understanding the importance of high and low factor scores. For the seven factors the desirable assessments give rise to:

- Learnt to 'seek help from others' – indicated by a low score.
- Becoming a 'considerate employer' – shown by a low score.
- Perseverance skills learnt – low score indicates this trait.
- Lessons from bad luck – low score indicates this outlook.
- Extensive and careful planner – low score indicates this approach.
- Discuss and mull over lessons from experience – low score indicates this is happening.
- Care and critiquing of others' suggestions – low score shows this occurs.

For the two measures based on the respondent's OII scores the differences between the top and bottom groups are largely highly significant. The exception is the 'care and critique' factor in the case of the productivity-based OII score. For the other two objective-weighted success measures, as will be noted, the significance levels are not as clear (other than the 'extensive and careful planner' and 'discuss and mull over' factors) – not that the other factors' 'non-significance' levels are poor.

The other issue to note is that despite the high significance levels the actual differences between the top and bottom groups are not high relative to the ranges covered by each factor, indicating that many respondents tend to have factor values in the middle areas of the distributions.

Age and Formal Education Impacts

When the data is examined to see if age impacts on the results it is abundantly clear that a manager's decision expertise increases throughout the life of a manager, reinforcing the dynamic nature of informed intuition. On the other hand, when education level is related to expertise there is little by way of a significant correlation – this probably indicates useful experiences which provide a manager with food for thought have impacts on successful informed decision making relative to formal educational benefits.

General

All these studies make it clear many aspects of the variables correlated with informed intuitive decision success need to be considered when investing in learning with a view to improving a manager's decision making. Making efforts to improve aspects of personality are definitely worth carrying out possibly using group cognitive behaviour therapy (Conard and Matthews, 2008; Proudfoot *et al.* 2009), and certainly additional training in the various areas should be pursued and most certainly the use of critical thinking should be worked on when learning every lesson offered through experiences, which will need careful recording.

Table 7.10. Cash change percentage OII scores. Average factor scores for the experience traits relative to the top and bottom groupings of OII ratings based on cash change percentage OII scores.

Experience factor	Top 25%	Bottom 25%	Difference	Significance	Range
Seek help from others	0.497	–0.460	0.957	<0.001	–2.69–4.33
Considerate employer	0.161	–0.177	0.338	<0.001	–2.70–5.15
Perseverance skills learnt	0.186	–0.172	0.358	<0.001	–6.11–2.88
Lessons from bad luck	0.211	–0.259	0.470	<0.001	–3.12–4.17
Extensive and careful planner	0.461	–0.448	0.909	<0.001	–3.47–3.26
Discuss and mull over lessons from experience	0.505	–0.415	0.920	<0.001	–2.53–4.00
Care and critiquing of others' suggestions	0.037	–0.046	0.083	<0.001	–3.66–4.19

Table 7.11. Productivity OII scores. Average factor scores for the experience traits relative to the top and bottom groupings of OII ratings based on productivity OII scores.

Experience factor	Top 25%	Bottom 25%	Difference	Significance	Range
Seek help from others	0.570	–0.495	1.265	<0.001	–2.69–4.33
Considerate employer	0.178	–0.201	0.379	<0.001	–2.70–5.15
Perseverance skills learnt	0.175	–0.129	0.304	<0.001	–6.11–2.88
Lessons from bad luck	0.185	–0.246	0.431	<0.001	–3.12–4.17
Extensive and careful planner	0.438	–0.426	0.864	<0.001	–3.47–3.26
Discuss and mull over lessons from experience	0.507	–0.434	0.941	<0.001	–2.53–4.00
Care and critique of others' suggestions	0.021	–0.023	0.044	0.249	–3.66–4.19

Table 7.12. Objective-weighted cash change percentage. Average factor scores for the experience traits relative to the top and bottom groupings of OII ratings based on objective-weighted cash change percentage scores.

Experience factor	Top 25%	Bottom 25%	Difference	Significance	Range
Seek help from others	0.303	–0.378	0.681	<0.001	–2.69 – .33
Considerate employer	0.007	–0.056	0.063	0.302	–2.70–5.15
Perseverance skills learnt	0.051	–0.027	0.078	0.232	–6.11–2.88
Lessons from bad luck	0.004	–0.107	0.111	0.084	–3.12–4.17
Extensive and careful planner	0.201	–0.282	0.483	<0.001	–3.47–3.26
Discuss and mull over lessons from experience	0.688	–0.455	1.143	<0.001	–2.53–4.00
Care and critiquing of others' suggestions	0.033	–0.050	0.083	0.111	–3.66–4.19

Table 7.13. Objective-weighted productivity. Average factor scores for the experience traits relative to the top and bottom groupings of OII ratings based on objective-weighted productivity scores.

Experience factor	Top 25%	Bottom 25%	Difference	Significance	Range
Seek help from others	0.458	–0.404	0.862	<0.001	–2.69–4.33
Considerate employer	0.024	–0.082	0.106	0.095	–2.70–5.15
Perseverance skills learnt	0.164	–0.098	0.262	<0.001	–6.11–2.88
Lessons from bad luck	0.074	–0.152	0.226	<0.001	–3.12–4.17
Extensive and careful planner	0.251	–0.277	0.528	<0.001	–3.47–3.26
Discuss and mull over lessons from experience	0.859	–0.643	1.502	<0.001	–2.53–4.00
Care and critiquing of others' suggestions	0.049	–0.053	0.102	0.121	–3.66–4.19

Another factor worthy of thought is the impact of emotions on decision making. It is likely issues like anger, sympathy, depression and the like may well influence decisions taken. This means where strong emotional factors are not prevalent when a decision is taken, the manager's decision is probably more likely to be appropriate and thus decisions should be avoided when it is clear emotion may well influence the outcome. Effectively, emotion may cause bias. On the other hand, not everyone will react irrationally under strong emotions so the decision may be appropriate, depending on the circumstances.

However, the research available on the impact of emotions is still very much in the emerging stage so managers seeking to improve should continue to focus on any research that might be conducted in the future. Useful references covering emotional topics include John and Gross (2004), Pfister and Bohm (2008), Proudfoot *et al.* (2009), Lerner *et al.* (2015) and Hughes *et al.* (2020).

In that there is little doubt excellent intuitive skill is very important for managers of SME businesses, managers must make every effort to improve and inform their intuition at every opportunity. The analyses presented, no matter the correct contribution of each core construct, makes it clear they all matter. Indeed, each manager will find each factor will contribute differently to how 'well informed' their intuition is. After all, each manager will have a different genotype, parental influences and other unique characteristics and contributors. The upshot is a manager must work on all factors through extensive reading and online searching, discussions, reviews, comparisons and critiques. A little searching will often find helpful resources such as Nuthall (2010), covering the core skills in managerial ability (such as observation and anticipation). Furthermore, it is useful to take time in reflecting on decisions as much as possible to allow the automatic cognitive processes time to operate and review. There is evidence (Croskerry *et al.*, 2013) that the conditions prevailing at the time of the decision can influence the conclusion. Reflection can counter any introduced bias such as that caused by extreme tiredness, or perhaps untoward emotions.

It is also critical for managers, and many others, to make an effort to learn the 'scientific method'. This does not mean they should become scientists, though this might help, but that they should learn the skills of being critical investigatively. This is not a negative approach; on the contrary, it is an approach that logically analyses each situation to come up with the conclusion that best achieves the manager's objectives.

With time this critical approach becomes second nature. It becomes part of a manager's intuition. The 'scientific method' has four components:

- observation;
- speculation;
- testing; and
- generalization.

Like most assessments, it is clear observation starts the process. In this case of decision making it must initially involve specifying the problem and then collecting all the relevant information that enables coming to a decision. Speculation involves considering which of the possible alternative decision choices might become contenders. Then comes analysing each possibility, leading to an assessment of the most appropriate for the objectives held. Finally, a cognitive process leads to a generalized conclusion that can be stored for later use should the same, or a similar, situation appear. With practice, this whole process should become automatic and part of intuition.

Others believe in slightly different processes. The Oxford Dictionary Online talks about the scientific method as:

> consisting in systematic observation, measurement, and experiment, and the formulation, testing, and modification of hypotheses. Experiments are a procedure designed to test hypotheses. ... The process of the scientific method involves making conjectures (hypotheses), deriving predictions from them as logical consequences, and then carrying out experiments or empirical observations based on those predictions. A hypothesis is a conjecture, based on knowledge obtained while seeking answers to the question. The hypothesis might be very specific, or it might be broad.

> Scientists then test hypotheses by conducting experiments or studies. A scientific hypothesis must be falsifiable, implying that it is possible to identify a possible outcome of an experiment or observation that conflicts with predictions deduced from the hypothesis; otherwise, the hypothesis cannot be meaningfully tested.
>
> (https://www.oed.com/dictionary/scientific-method_n?tab=factsheet#23962743, accessed 9 September 2024)

In the manager's case it is most unlikely experiments are possible except in their mind's eye, though it is not unknown for managers to carry out simple experiments such as trying different advertising levels. But in most cases the procedure is a mind-and-desk process supported by the experiments of others, such as the nearest research group or station.

But above all, every step requires 'cynicism' or critical thinking. That is, everything is examined for its believability in contrast to blind acceptance of the information collected and recounted by others. Each step needs to be checked so the decision maker is convinced the data and knowledge is as accurate as it is humanly possible to obtain. In his book, Hogarth (2001) talks about the 'scientific method' and concludes successful intuition should be called 'educated intuition'. This is hard to argue against and indeed the term 'informed intuition' has been used in this book which is clearly very similar.

Over the years various experiments have been carried out on methods of improving intuition. Similarly, many ideas have been voiced to the same ends. Some of this work is reviewed by Nuthall and Old (2018). There are two conclusions that emerge.

The first approach involves managers and professionals getting together to review each manager's decisions. The group then brings to bear their collective wisdom on the efficacy of the decision. This process can also be used prior to the decision. The group size can vary from two or three, to many more up to, say, six or seven. More becomes unwieldy. The group might also be a family group which might meet regularly anyway. This process is fully explained in Nuthall (2016).

Another approach has the manager operating more on her or his own. The manager keeps a detailed diary of decisions taken which gives:

- the details of the decision situation and the information gleaned;
- the logic behind the thinking used in coming to the decision and the processes used;
- as the decision and its impacts unfold, the explanation and the outcomes observed are kept up to date;
- once the impacts have played out fully and physical and financial outcome data is available, the manager undertakes and records a review of the cause, effect and outcome factors; which
- eventually enables coming to a conclusion on whether this was the correct decision, and if not, what is believed would have been correct.

As each decision is reviewed in this way the manager's intuition will improve. Eventually the manager may be able to minimize the use of a diary and associated reviews as confidence and success evolve. Some decisions will be major, such as buying new machinery or introducing a new product, others will be less important such as changing the brand of a paint being sprayed on a product, but in all cases it is useful to record the decision and write down reflections.

Of course, some users will find both these suggestions valuable when operated together. Group sessions will clearly benefit from sharing the diary.

In some cases formal courses will also provide value. However, they may not be available locally and time might be an issue. One course topic which would always be valuable is production economics. And many relevant courses are available online, and even in textbooks! A difficulty for managers is, of course, finding high-quality time to study. It takes determination. That is why group work and diary keeping may be more practical.

But in the case of management students, the situation is quite different.

These suggested approaches to improving intuition dovetail into the more extensive discussion in Chapter 10 covering improvements to managerial ability in general. The two discussions form a whole.

Concluding Comments

Part of improving a manager's intuition must involve assessing biases. Any manager that does not achieve their objectives will find it is due to either unrealistic objective expectations and requirements, untoward risk and uncertainty impacts, or a biased decision and plan implementation process. One of the first places to look for improvement, other than through the training systems suggested above, is the existence of biases. Chapter 6 contains a review of possible biases which should be examined as part of a manager's self-review system. The manager might well enlist the help of others. Also contained in Chapter 6 are suggestions for methods in finding the likely biases occurring. Once isolated, correction and repair can begin.

The conclusions reached on intuition further enhance understanding the concept of human capital, and also social capital. For example, for social capital the results have shown the importance of community links and consultations, and, for human capital, the importance of such processes as self-critiquing, and many others.

While only briefly mentioned previously, the importance of genotype in intuition is an interesting question. How important is birth relative to training and experience? To further explore this issue it would be necessary to enlist the help of identical twins, both of whom ended up as managers but in different situations. It will have been noticed the regression on productive efficiency explained 70% of the variance – does this suggest the remaining 30% of variation is due to genetic effects? If this is the case it demonstrates the importance of a manager spending time specifically learning from experience and training. There are many studies demonstrating the benefits of training and education (e.g. Xayavong *et al.*, 2015).

Perhaps clues on the importance of genotype come from the research on personality using identical twins. It has been shown the heritability of extraversion is 53%, agreeableness 41%, and lesser amounts for the remaining traits. A figure of 30% plus for intuition is accordingly within the realms of possibility. But whatever the case, a manager will find it useful to mentally test each of the conclusions arrived at through cognitive activity and even using mental, and calculator, based 'check-ups'. For example, a manager of an animal-based production system might come to an intuition-based decision to increase feeding levels, but with a few sums the maintenance and production requirements can be calculated and checked against the likely intake.

Future work on intuition might well explore tests to check proficiency. Wagner and Sternberg (1987) talk about the idea of getting accepted intuitive experts to describe decision scenarios and what their decision would be. The manager's response to the conditions can be checked against the expert's conclusion with the degree of differences providing a score. There could be a range of tests emphasizing different forms of decision scenarios. Certainly, as borne out by the results, the correlation between intelligence and intuitive skill is not great, meaning standard intelligent tests are of little use as proxies.

The last word on intuition is probably a long way off. The research reported here needs replicating and repeating in other situations to ensure the results are undoubtedly robust. Furthermore, additional variables should probably be added to the analyses – examples include factors such as a manager's willingness to learn, their learning style and the nature of each experience as perhaps extreme situations might be more valuable than others. A researcher interested in further work on intuition might well find a book (Sinclair, 2014) on research methods for intuition a good starting point.

Each manager's experience set will be somewhat random, and to complicate this is each manager's ability to learn from each occasion. The other important factor is the dynamic nature of business and the use and learning of intuition – the particular sequence experienced may be important. Despite this, having an open mind to future developments will be important, but nevertheless the variables shown to be important in the static case are more than likely important in the real-life dynamic case. Anyone interested in the dynamic case could well reflect on Fig. 7.5.

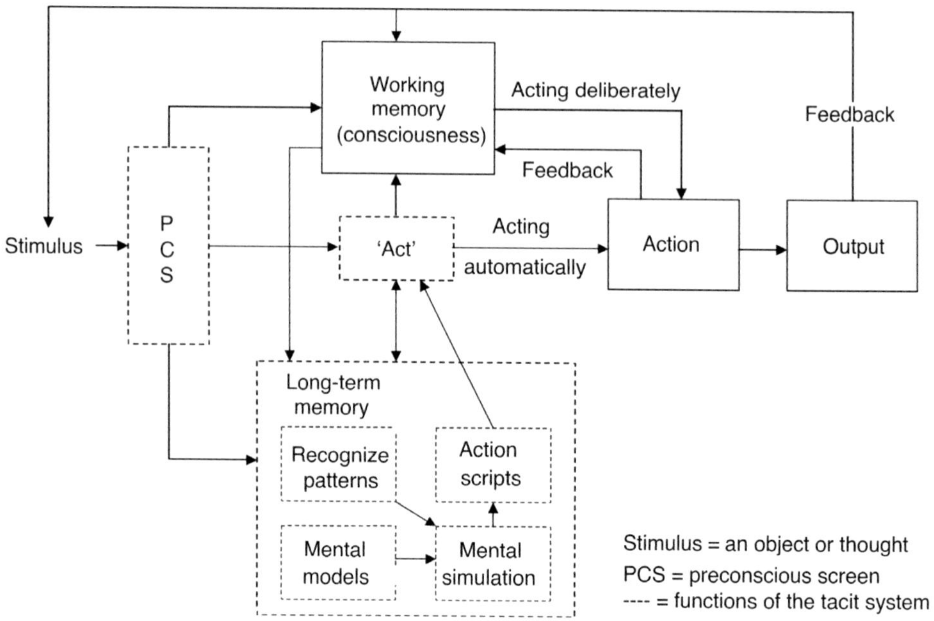

Fig. 7.5. The tacit and deliberate systems of human information processing (after Hogarth, 2001; Klein *et al.*, 2005). (From Ohlmer, 2007.)

What is clear, however, is that if a manager wishes to improve her or his intuitive skills a special effort is required with time being set aside for homework. The tasks must be set carefully and logically to cover the contributing components to intuition and completed with all details recorded. Improvement is seldom serendipitous.

8 The Influence of Managers' Personal Characteristics on a Range of Issues in Management

Abstract

The 'human factor' has a major influence in a whole range of SME decision issues. This chapter reviews a range of studies exploring the impacts of managers', and other personnel's, characteristics on important issues. The first study examines business growth, finding, for example, that older managers tend to be 'expanders' with low anxiety and a high belief in their control (Locus of Control) level. Their objectives are also important. Another study examines ownership and governance questions using a quantifiable model. Manager personality (management style) and objectives strongly influence the choices. Business succession is another problem area modelled to explain the reasons why managers tend to be 'laggards' (slow to organize succession). Reasons include family size, personality and health, with other 'barriers' also discussed. The final study covers manager anxiety. The model developed explains business, debt and family anxiety, and others too, again showing the influence of personality. Overall, the 'human factor' is all pervading.

Introduction

The thesis behind this book is that the human factor has an enormous influence on the life and times of any SME business. The information presented in the previous chapters makes this clear in a general sense. This chapter contains material that puts more flesh on the assertion by reviewing a number of studies covering a sample of aspects impacting on businesses. It is also important to realize that the 'human factor' is part and parcel of all *Homo sapiens* involved in the life of businesses right from the new labourer through to the owners who may or may not directly contribute to the day-to-day running of the business.

In this chapter it is the manager whose human factor is brought further to the fore, but in so doing it should be remembered that her or his interactions with all the other humans involved in a business may be influenced by the characteristics of each and every one of the participants. A business operates not only by the planning decisions taken by the humans, but also by how successfully they carry out what has been decided. And the whole process is dynamic as people, risk and uncertainty unfold.

In that every human is unique, every business will be largely unique in its operations. This is where 'manager individualism' comes in. Despite this, it is possible to provide some generalizations about human characteristics and their impact on various issues. This can only be for a sample of issues as it is the obvious truth that every outcome in SME, and other, businesses is influenced by the human factor one way or another. That is why, of course, it is so important to study the human factor, and how it might be influenced for the better.

The particular issues chosen to be discussed include the growth of small businesses, ownership system considerations, business succession problems, and manager and business owner anxiety. There is no common link between these topics other than they are all influenced by the human factor. Detailed information is available about each, thus accounting for their selection, and they do cover a wide range of business and production issues. Clearly many others could have been chosen, but the coverage will reinforce the wide-ranging impact of the human factor.

Another issue that might have been selected is a manager's risk attitude, which is heavily influenced by the manager's personality, parenting

DOI: 10.1079/9781800627246.0008

and other people's influences not to mention the impact of the business's environment both physical and market wise. However, an online search will provide details of many already available studies on utility and risk attitudes.

A further example is provided by Austin *et al.* (2001) showing how personality and cognitive ability impact on production orientation, environmental attitudes and diversification (risk influence). Somewhat similarly, Crase and Mayberry (2003) showed the importance of personality, objectives, attitudes and intentions on influencing the adoption of conservation practices. Furthermore, in reviewing the literature, McElwee (2006) stressed that it appeared successful entrepreneurship was influenced by personality-related issues such as the manager's locus of control, problem-solving abilities and social skills.

The examples in the literature even extend to the influence of personality on the voluntary vaccination of animals (Sok *et al.*, 2016), and on technology adoption through what was labelled a 'family bargaining process' in a less-developed agricultural country using the theory of planned behaviour (Sambodo and Nuthall, 2010).

The chapter proceeds by presenting details of each of the cases highlighted and ends with concluding comments.

Business Growth

Introduction

Business growth is pivotal for many producers and managers for a range of reasons. In some cases the business is too small to be economically sustainable, and in others the manager might simply wish to expand as it is in the manager's psyche to constantly seek growth. For still others, while colleagues, family and friends might suggest they should expand, they may in fact be content to exist within their current purview and correspondingly adjust their lives to cope with what they have.

The study reported here (Westbrooke and Nuthall, 2017) works on understanding the reasons individual managers take one of the courses outlined above. Why do some remain static, and others take quite the opposite path? Figure 8.1 represents the possibilities and the likely influencing factors. It is hypothesized the manager's personal characteristics together with details of the existing business situation heavily influence the decisions taken. In this process it is assumed the business family, if any, will influence the manager's actions through their influence on the manager's attitudes.

It will be noted a manager also has the option to sell the business which may or may not lead to another purchase that, hopefully, is in a better situation. In some cases selling might be due to retirement, or perhaps ill health, with any offspring, or other heirs, not being interested in taking over the business.

Seldom can managers stand still due to the constant cost price squeeze they face. The price squeeze has been going on, on average, for decades. If the manager's current cash surplus is large then perhaps they have a few years' grace, but in the end action is necessary. This action might be to intensify through a range of options including, perhaps, introducing new products or production methods, expanding the stock of land held or, as noted, the exact opposite through selling.

This is all expressed in Fig. 8.1. The question is what influences a manager to be a 'consolidator' or 'expander' (develop), and if an expander, what influences the paths chosen? The third option is to be a 'retractor' (sell/retire).

The data

To discover some of these issues, a phone survey was undertaken of smaller businesses that were randomly selected from both members of a small business association and electoral rolls in which each person's occupation was listed. The number of successfully completed interviews was 346 and to this information some common questions from two other earlier mail surveys were added to provide comparisons, particularly in looking at size differences. To qualify as a smaller business in the study, the number production units had to be less than 400 which is, of course, a random cut-off but needs to be seen in the context that the average national business of the particular type has 419 production units despite over 11% having more than 1000 units.

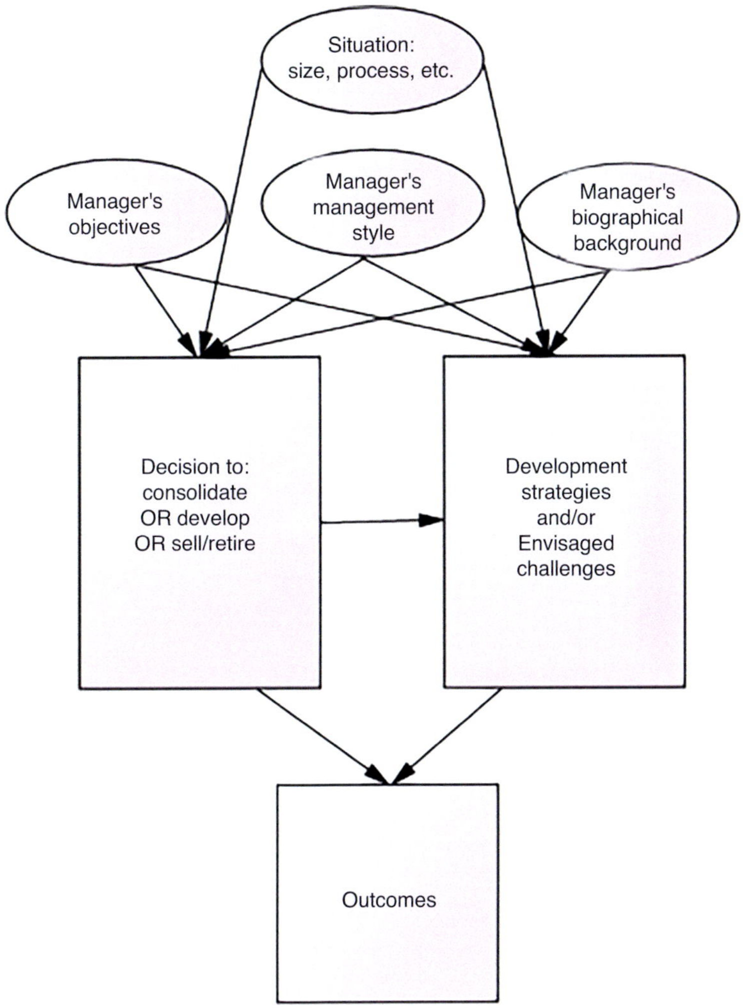

Fig. 8.1. A schematic of the components of a manager's development choices. (From Westbrooke and Nuthall, 2017.)

The questionnaires sought information on a range of both business and personal variables which were then used in the analyses described below. The managers were also asked information on the history of their business size to allow comparisons between the expanders and the consolidators.

The analysis

The comparison showed the expanders were significantly older than the others, which is to be expected in part as they have had more time to develop. Interestingly, there was a significant difference over their place of birth – the expanders tended to have been born in urban areas, suggesting perhaps they were less steeped in traditional rural ways. However, despite this, the expanders on average were slightly less educated by a third of a point on a 1 (primary) to 5 (postgraduate) scale, and no doubt due to the age comparison, the expanders had a slightly higher equity (6.5% significant difference) and correspondingly better resources for expansion.

Of greater significance, however, from a management style point of view, was that the expanders were:

- Significantly more tolerant of mistakes made by their employees and the contractors employed. This is a factor related to personality.
- Similarly, the expanders tended to be lower on the anxiety scale.
- And, very importantly, they had a stronger belief in their ability to control outcomes (negative to 'problems are due to factors I can't control') with respect to their locus of control.
- Also relevant is that their objectives were not significantly different. Thus, other than the age, education and equity differences, their personal characteristics were the main factors in the different courses taken.

When it came to comparing policies of the managers (the right-hand rectangle in Fig. 8.1), the sample was clustered into six groups based on their objectives and management style variables. Each group had approximately the same number of managers, which was the objective when selecting six groups. Each group had distinctly different objectives and personalities (management styles). The average 10-year development plan for each group was then compared, with the differences tested for significance.

The results showed the following development (change) options were defined by the managers' objectives and personality:

- selling outright;
- selling the business to buy bigger;
- increase existing business size;
- transferring the business to the manager's heirs;
- doing the majority of business work yourself;
- investing in work-saving technology;
- increase production by more than 10%; and
- investing in an additional business.

And for the challenges the managers perceived in any development process, the factors significantly different between the groups were:

- the level of cash income over the last 4 years; and
- expected future cash returns.

There are no surprises in this list. Issues that were not significantly different between the groups were their level of knowledge, the level of risk involved and the amount of capital/debt required. In addition, the education and rurality of their birth were significantly different across the groups.

Overall it is clear many of the important factors in the development problem are defined by the personal features of the managers. This has many implications for extension, manager improvement and policy work.

To reinforce the conclusions, the managers' attitudes to the development challenges were used to factorize the managers' approaches into three factors (a sort of clustering but based on the policies the managers were planning on using) and the three groups were given names based on their policy groupings. These were:

- 'finance and risk' (cash return, debt and risk levels were dominant issues);
- 'labour and knowledge' (staff management ability, finding suitable staff and knowledge featured); and
- 'technology and environment' (environmental regulations, and a lack of knowledge thereof).

Similarly the 10-year plan options were factorized to give four factors:

- 'all-round expander' (the individual variables that featured were increase business size, increase production, invest in another business, business technology);
- 'employer' (the variables featuring were employ staff and reduce debt);
- 'off-business income earner' (important variable was the desire to earn at least 20% 'off business'); and
- 'business trader' (with variables 'sell business', sell to go larger).

The results were then subjected to a regression analysis to assess the factors giving rise to these attitudes.

For the 10 year plan options, the 'all-round expander' equation had an R^2 of 0.375 (with a significance probability of 0.000), the 'employer' an R^2 of 0.547 (0.000), the 'off-business income earner' an R^2 of 0.275 (0.000) and the 'business trader' an R^2 of 0.327 (0.000).

In the case of the first two factors the managers' objectives and managerial style were particularly important in determining these attitudes, but less so for the 'off-business income earner' and 'business trader'. The 'logician', 'consulter', 'conscientious planner' and 'benign manager' were all important determining variables as components of management style, and for the manager objectives, 'risk remover' and 'community supporter' components were important. In the case of the other two factors, the objective components 'balanced', 'way of life' and 'family supporter' featured. Also important were variables such as the level of output produced, equity and the managers' age.

For the challenges, the managers' objectives had considerable influence. The 'balanced' factor influenced all the challenges, as did the 'way of life' and 'reluctant manager' aspects of their objectives. For the management style components it was really only for the 'finance and risk' challenge that there was much impact. The 'thoughtful creator', the 'benign manager', the 'correctness seeker' and the 'logician' were all personality components giving rise to the 'finance and risk' factor.

Overall, if the coefficients are summed, the importance of the manager's objectives varied from 16% through to 51% in determining the 10 year plans (depending on which specific development factor). For the management style constituents their influence on the plans ranged from 3% through to 50%. These percentages relate to 27 to 40% for the other biographical factors, and 6 to 17% for the production variables.

The importance of the personological (objectives and management style) factors is clear. However, it must also be noted that the R^2 values show that many other factors must also be involved in the explanations.

Comparison of the managers on large as against smaller businesses

If the primary hypothesis that the human factor does influence many issues is correct, you would expect the nature of managers in small businesses to be different from those operating larger businesses. To test this two, earlier surveys were used to extract information on larger businesses for comparative purposes. Table 8.1 contains the results, giving the statistics for each and the significance levels of the comparisons (columns 3 and 5). The information covers the differences in factors for the objectives, management style, locus of control as well as biographical/production data.

Overall, there seems largely to be conformity in the signs on the coefficients, and, mainly, consistency in the two large business surveys. It will be noticed the production units per labour unit is higher in the larger businesses, as is the output per labour unit (kg), indicating, probably, the impact of business size on efficiency. While the difference in education levels between the small businesses and one of the large business surveys is significant, it is not great, with the managers of the smaller businesses having a higher level of formal education.

However, when it comes to the objectives, there are many clear and significant differences. A legitimate question is 'what comes first?'– the small business or the core objectives? You can imagine altering your objectives to suit what is actually possible over the years given the resources on hand. Probably, in the end, a bit of both occurs with some of the impacts of the objectives restricting a small manager's urge to grow the size of the business.

When it comes to the managers' management style, there are clear differences between the small and larger businesses. An interesting question, given the reasonable heritability of personality, is whether management style has been similar through the generations living on any one business? And has this impacted on the business as it has been passed on from one generation to the next? The data does not allow a conclusion here, but, for many of the smaller businesses' managers, the current manager's management style has probably, in part, led to a lack of ambition to expand.

Messages

As far as managers of small businesses are concerned, it is clear they tend to be inherently

Table 8.1. A comparison between small managers' objectives, management styles and biographical/ production information relative to larger business's managers. Average factor scores for each group with the larger businesses coming from survey '06 and survey '13. (From Westbrooke and Nuthall, 2017.)

		Survey 1		Survey 2	
Objective/Style/ Variable[a]	Smaller businesses	Large businesses	*t*-Test significance probability (column 1 versus column 2)	Large businesses	*t*-Test significance probability (column 1 versus column 4)
	(1)	(2)	(3)	(4)	(5)
Objectives					
Obj. 1 – balanced	0.999	0.089	0.000	0.034	0.000
Obj. 2 – risk remover	−0.193	−0.007	0.004	−0.194	0.978
Obj. 3 – way of life	−0.628	−0.116	0.000	−0.157	0.000
Obj. 4 – reluctant manager	0.567	−0.099	0.000	−0.227	0.000
Obj. 5 – community supporter	0.174	0.401	0.007	0.212	0.588
Obj. 6 – family supporter	0.282	−0.030	0.001	−0.085	0.000
Management style					
Style 1 – consultative logician community	−0.255	−0.178	0.388	−0.272	0.819
Style 2 – correctness seeker	0.185	−0.062	0.001	−0.072	0.000
Style 3 – consultative logician family and friends	0.198	−0.082	0.000	−0.087	0.000
Style 4 – conscientious planner	0.283	0.024	0.000	−0.038	0.000
Style 5 – thoughtful creator	−0.293	−0.136	0.031	0.022	0.000
Style 6 – benign manager	−0.410	−0.093	0.000	−0.076	0.000
Locus of control (%)	67.22	67.55	0.600	68.17	0.008
Biographical/ production data					
Production units per labour unit	59.0	81.4	0.000	71.5	0.000

Continued

Table 8.1. Continued

Objective/Style/ Variable[a]	Smaller businesses	Survey 1 Large businesses	*t*-Test significance probability (column 1 versus column 2)	Survey 2 Large businesses	*t*-Test significance probability (column 1 versus column 4)
Total labour units	1.83	3.77	0.000	6.20	0.000
Output per labour unit (kg)	51,290	73,323	0.000	80,732	0.000
Manager age code (1 = '< 30 years' to 5 = '> 60 years')	2.87	3.43	0.000	3.59	0.000
Education code (1 = 'secondary' to 5 = 'postgraduate')	3.29	3.01	0.000	3.65	0.574

[a]See Chapter 7 for definitions of the objective and management style factors. Due to the scoring system, with one meaning tending to the description of the objective or style, and five the opposite, a lower score means true. The scores are negative in many cases as they are the factor estimates based on the combinations of the core questions.

different from managers of larger businesses, and when it comes to their wishes and methods associated with growth, they are more than likely impacted by the human factor. This again reinforces the importance of studying the businesses when working with these managers, and for the managers themselves, it is clear they may find working on changing their attitudes can provide many benefits. The results also emphasize the need for policy workers to consider manager personality.

It was also interesting that a significant number of the smaller businesses were content to remain small despite the cost price squeeze. They worked at improving efficiency and some sought off-business income from various sources, while others ventured into a new enterprise such as tourism. It might be concluded 'where there was a will there was a way', and part of this was reflected in their efforts to reduce debt whenever the slightest cash surplus was produced. This improved the cash surplus and equity and consequently provided a buffer for future difficulties. And it all stemmed from their objectives and personalities.

Another powerful factor was the business history – if it had been in the family several generations, the current incumbent felt the need to maintain the tradition.

Ownership Systems

Background

For most businesses, their ownership and governance systems were probably handed down from their forebears. Despite this, it is possible for a manager to change the ownership and governance system at any stage including at the initial transfer. Similarly, when a business is newly purchased there is the chance to decide what the manager, and/or owners, believe is the most appropriate structure. Whatever the case, it is important to understand the influences being brought to bear on the decisions made over the ownership structure. This section addresses this problem with the notion that it is likely the 'personological' variables will be associated with the current ownership structure on any business.

To assess this situation the relevant literature was reviewed and a set of data from some 800 plus businesses obtained and analysed. The questionnaire used and details of the basic analyses are given in Nuthall and Old (2014), which is available online.

In the Western world, and in many other places too, a wide range of ownership and governance systems is found. Furthermore, usually ownership and governance systems go together in that, for example, if the business is owned by one person who is the manager and a worker, it is usual for both governance and day-to-day operations to be controlled by this owner/operator.

While this owner/operator system is most common worldwide, it is also possible to have partnership arrangements covering ownership and governance. How much of each is conducted by each partner/s varies. For example, if the partnership is a husband–wife situation, it is not uncommon for the husband to make most of the decisions and to implement the programme, though increasingly consultation will occur. Of course, it is also possible for the arrangement to be the other way with the wife (female partner) being the main operator.

People other than family may also be involved in partnerships. One example is an equity partner – this is where one or more partners are 'sleeping' when it comes to business operations but provide capital to help purchase and run the business. In any one case, the detailed partnership agreement determines how profits are distributed, and in many cases the managing partner may have a salary paid.

Another possibility involves what are called trusts. A trust is usually a recognized legal entity which owns assets and has trustees controlling decisions. In many cases trusts can be part of the set-ups already mentioned with, perhaps, a trust owning some of the assets. In many jurisdictions this has the advantage of limiting liability to the trust's assets so that a manager's personal assets are removed from the liability mix. A trust can be set up to benefit a range of beneficiaries such as a business couple's children.

Finally, there is the company (or corporation) ownership and governance system. Generally a company can be either family based, or open to the general public (i.e. a public company in contrast to a family (private) company). The company has owners through their share, or stock, holding. Each company will have a defined number of shares with various proportions owned by family members in a private company, and by members of the public in public companies. Voting rights usually relate to the number of shares held. The big advantage of companies is that liability is limited to the company assets, isolating the owners' personal assets and, accordingly, involving less personal risk. And some would say companies have financing advantages, and also the advantage of many checks and balances on decisions. Some observers might not agree that these supposed advantages are real.

Each company has a board of directors responsible for governance whereas the manager appointed by the board is responsible for the day-to-day operations even if overseen by the board. The downside of having a board is the costs involved with directors' fees and other accounting and legal costs. As noted, the upside of a board is the checks and balances provided and the additional expertise available assuming each board member has abilities in a range of different areas.

There is also a compromise possibility. Some businesses have an advisory committee in place of a board which means a legal entity complication does not exist compared with a formal board. A committee, however, may have associated costs.

It is necessary to be clear on the various alternatives to allow comparisons. Many will not be a simple company, trust, partnership or owner/operator arrangement. For example, a mixture of partnerships and trusts is possible as are a private company and trusts, and so on. Besides the general comments made, the details and legal responsibilities will depend on the jurisdiction under which each business operates.

Another factor is the tax system, and the associated tax rates, applied to each kind of entity. Variable rates between structures may be a reason for having, say, trusts compared to owner/operators.

An important consideration in a choice of system is the complexity of each business domain. Biological systems, for example, continue 24 hours per day over 365 days each year. It is never possible to close the 'factory' door like in many straightforward urban-based businesses. This means a manager and family, or their representatives, living either at, or in close proximity to, the business are available night and day to immediately attend to issues requiring attention. Waiting for normal work start times only to discover a major problem means efficiency slips well down on what is possible. With primary production in particular, the old saying 'a stitch in time saves nine' is very relevant. The relationship between people working on a complex, technically based business and the technology is critical and probably relates to an individual's personality. Some people are empathetic to, and understand, the basis of the technology and machinery, others are not.

Furthermore, in considering ownership and governance, many writers discuss the relationship between governance and operational management with its daily responsibilities. The proponents of company structure talk about this separation between governance and operational responsibilities being an advantage and certainly an issue that must be protected. But in reality, on an SME business, there is seldom a clear distinction, with people intimately involved in a business's life also being a party to all decisions whether they might be classed governance or operational. One influences the other. Whatever the arrangements, decisions must be made about all factors from the bottom up to the final marketing of products and issues such as financing. It is also relevant that a business usually involves a relatively small number of workers who work closely together in contrast to a larger public company operation where personal relationships are less significant.

Having briefly outlined the ownership options, the discussion continues by considering the factors that might influence the decision on the appropriate ownership and governance structure. Analyses of the data collected are presented, and conclusions on the factors which tend to dominate the choice of set-up, at least for the particular set of managers answering the survey, are presented and discussed.

A generalized model of the variables likely to impact on the choice of ownership and governance

In the beginning, according to the first records of organized production in countries like Egypt and the Middle East, businesses were largely family oriented. Before the beginning, hunting and gathering were dominant though, no doubt, the two systems merged and crossed over. Even where communal land situations existed families were often allocated particular segments which overcame the problem of the commons.

Eventually communities started to develop traditions to control the ownership of land rights and their governance. Laws, either written or not, appeared, defining what managers could, and could not, do. Eventually, the various systems described appeared. Despite all the possible alternatives, family business (largely owner/operator) still dominates the system today.

Nuthall and Old (2017) review this situation and note that in primary-production-related businesses:

- of New Zealand businesses 91% are clearly family oriented;
- 84% of Canadian businesses are owner/operator or partnerships; and
- in Europe 97% are family based.

With the constant decrease in the number of businesses in each country as amalgamations occur, it might be imagined there would be an increasing interest in corporate systems. Undoubtedly this will be the case in some eras, but it must be remembered that returns in many SMEs are seldom as high as returns in large businesses due to the non-monetary benefits business families obtain from being managers. This gives a demand for SMEs above what purely economic returns would stimulate.

Another trend that might promote company systems is multiple business ownership. For example:

- in New Zealand, the average number of primary production businesses held by one family is 1.75 (Nuthall and Old, 2014);
- in the USA, 21% of holdings are shared by more than one nuclear family; and
- in Canada, the equivalent figure is 16%.

Given all these trends it is important to understand their importance, together with all other drivers of the choice of ownership/governance systems. Should it be owner/operator, partnership or corporate? In this choice it must also be remembered the family business can be quite resilient relative to companies which need to perform financially otherwise investors will attempt to pull out their money. Most 'arms-length' investors do not get the benefits of living on businesses, so their main interest is the direct monetary return.

SME business families are very different to corporate society and often accept the need to be stringent in financial downturns and are resourceful in finding other incomes. They often survive quite poor periods and recover to reap 'upturn' profits, many of which are invested 'off business' to provide a buffer for the inevitable next downturn. Difficulties occur, however, when a new manager, heavily indebted, initially strikes a downturn. Luck is important.

Family businesses have many other advantages that are reviewed by Nuthall and Old (2017), as are the benefits of corporates, and, by association, partnerships.

To quantitatively compare the systems it is necessary to evaluate the variables likely to impact on the choice of system. Considering the discussion so far, it is possible to list the logical and likely influences. These are portrayed in Fig. 8.2. In the comparison, an existing manager, and similarly a manager investing in a business for the first time, must assess:

- the purpose of being a manager;
- the rewards available under each system;
- the resource implications of each;
- the relative risks involved under each structure; and
- the ease of succession arrangements.

A manager's objectives must dominate any analysis, assuming they have a large personal investment, as should their personal attributes covering such factors as the value of having freedom in a business which is as much related to personality as is the manager's risk attitude. Being your own supervisor is very important to many managers and is part of a manager's independent spirit. Following on are such factors as costs and financing, tax and succession, and the use of advisors, which must all be assessed in the decision as shown in Fig. 8.2. The process culminates in choosing from the continuum represented by the ellipses at the bottom of Fig. 8.2. While not specifically mentioned, sustainability is probably important to managers who may well be interested in ensuring their offspring still have a viable set of assets.

It will also be noticed managerial ability, and intuitive ability, must also be factors for clearly if a manager is really knowledgeable and has high ability, then the benefits of a corporate system are largely negated especially where financing is not a problem. In this study a manager's managerial ability was obtained from using a predictive equation developed by Nuthall (2009).

The results from the analyses

The research presented relies on the managers of each business being the dominant force no matter the ownership and governance arrangement. As shown by the data in Table 8.2, this is, surprisingly, clearly the case for most situations. Accordingly the managers were asked to provide their details for variables such as their management style, objectives, number of children and the like, which were all used in the analysis.

Friedman's two-way analysis shows there are no significant differences between the rows (p=0.459) but the columns are significantly different at 5% (p=0.035).

To decide which variables relate to each of the basic ownership forms, the average value of a wide range of statistics was calculated for the businesses using each ownership system. The results are contained in Table 8.3 together with the significance probability of the differences.

Some variables are much the same no matter the ownership system. Examples include the net assets per labour unit (including the manager), the profit per labour unit, the money spent on advice, the knowledge of governance systems, and, surprisingly, the manager's ability.

Differences occur in the age of the managers with company managers being younger, which is partly a function of company structures only being suggested by lawyers, accountants and advisers in recent years. Similarly, the managers of company businesses tend to have higher education probably in response to education creep. Little can be concluded for the proxy of intelligence, average grades, even though significantly different.

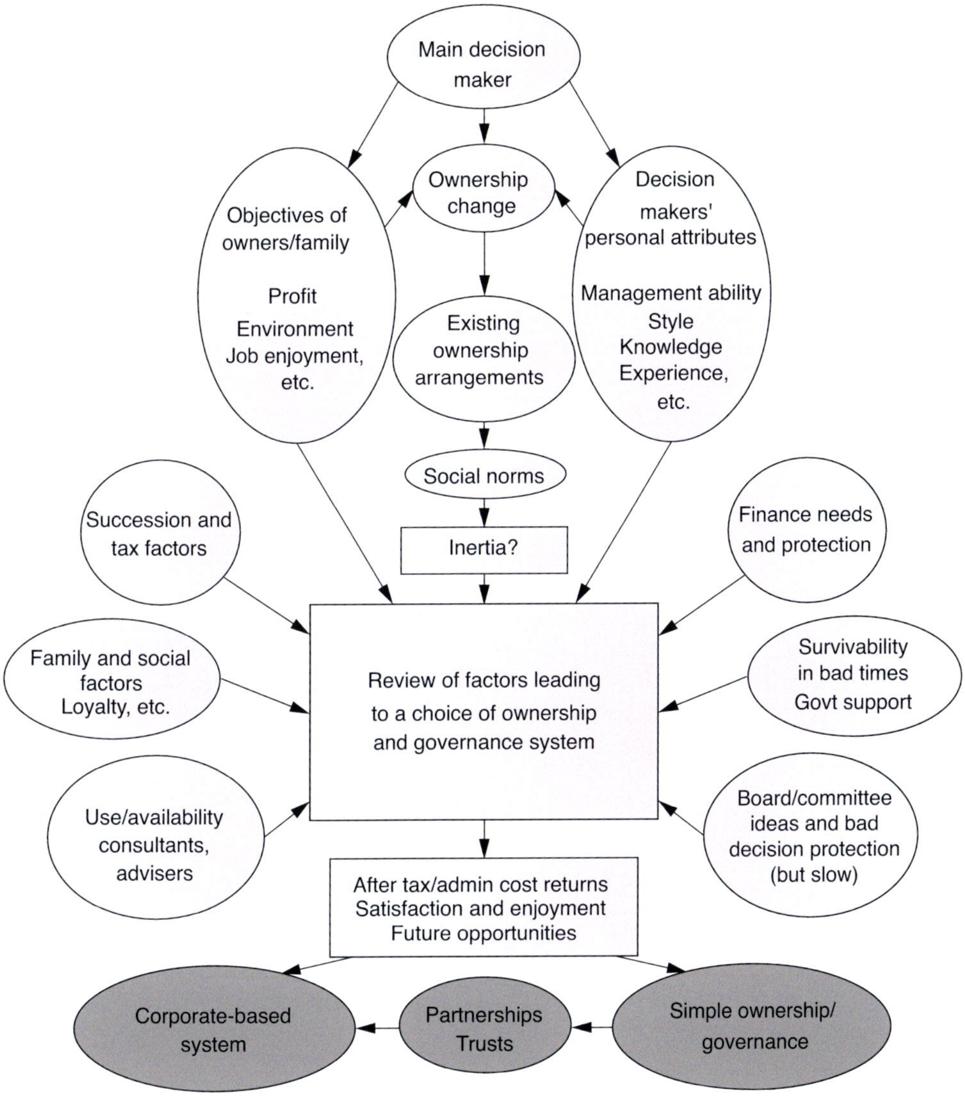

Fig. 8.2. A model of factors influencing a business's ownership and governance systems. (From Nuthall and Old, 2017.)

It is also clear most managers are male with company business managers having smaller families, fewer years in the business, larger businesses to manage (number of labour units) and each company has more individual businesses in the mix.

Most of this is to be expected, but what is very striking is the differences in the management style variables, or personality. Company managers, who are mainly part owners given the predominance of family companies, tend to be more anxious than the others, but also more conscientious, more agreeable, exhibit greater openness and tend to be more extravert (in general their scores are lower, given the scoring system as noted in Table 8.3 footnotes).

And just as importantly, the objectives of managers of the largely family companies are significantly different. They tend to be more

Table 8.2. Percentage of managers making decisions using various support arrangements under different operating/governance scenarios and investment levels. The first sub-column under each heading is for '*strategic and long-term policy*' (St.) and the second '*tactical and/or short-term questions including day-to-day decisions*' (Ta.). Details of each main column are given in the footnotes. The data was rounded to the nearest integer. (From Nuthall and Old, 2017.)

Asset range	Make all with help (1)		Confer consult (2)		Com't'ee (3)		Board (4)		Partnership (5)		Sole decider (6)		Trustees (7)	
(NZ$)	St.	Ta.	St.	Ta.	St.	Ta.	St.	Ta.	St.	Ta.	St.	Ta.	St.	Ta.
< 10 million	78	69	35	45	5	7	3	3	55	54	61	51	27	23
> 10 million	59	53	54	52	20	19	22	18	62	50	30	21	36	30
> 15 million	54	55	62	61	28	19	24	25	55	46	26	22	29	20

1) '(1) 'Make all' = make all decisions but with advice from family/friends/colleagues; (2) 'Confer consult' = frequently confer and take advice from a professional consultant; (3) 'Com't'ee' = often have committee of lay and professional people to help using formal meetings; (4) 'Board' = have board of directors that frequently meet and has the final say; (5) 'Partnership' = as a partnership we make most decisions; (6) 'Sole decider' = make decisions without discussions with others; (7) 'Trustees' = business is owned at least in part by a trust and you consult the trustees.

interested in lifestyle and leisure, ensuring work is enjoyable, achieving a good monetary return and growth though the risk aspect of their objectives is indeterminate given the significance probability. However, they do involve others in obtaining risk advice, which is all very logical. And they believe they know about finances in some detail.

It will also be noted for many variables the figures for partnerships lie between sole traders and company arrangements. However, this is not the case for the percentage of decisions made by the manager where, for companies, it is higher than for both the other forms of ownership.

Overall, it is clear all the personal characteristics of company managers are what you might expect a progressive business manager to exhibit. Furthermore, when the productive efficiency of each group was compared it was found company businesses had a coefficient of 1.03 relative to sole traders coming in at 0.99 and partnership systems at 0.98. The differences are not major.

Messages

This comparison does not prove it is the differences in the manager's personal characteristics that determine ownership form, but it does strongly suggest this conclusion – through, probably, a business owner's personality and objectives which lead to an interest in changing to what they might regard as having advantages, as informed, no doubt, by professional advisers.

To further explore the issue, a logistic regression was calculated to provide a predictive equation for each ownership form. The predictive power of each equation was 74.2% for sole traders, 60.7% for partnerships and 91.4% for companies. But of more importance is that when the personality coefficients were averaged, for sole traders their influence was 0.97, for partnerships 1.12 and for companies 1.28. The equivalent variables for the objectives were 0.96, 1.23 and 2.05. To interpret these logistic regression figures some will remember from their statistical studies that if the figure is over 1.0 it means the probability of the business being the particular ownership form is higher, and below 1.0 it is lower. It is clear from these figures how the personal variables are influential. For the business characteristic variables, their influence was minor in comparison.

Finally, the average return on capital was calculated for each form, with sole traders achieving 3.72%, 3.24% for partnerships and 3.16% for companies. Other than legal liability questions, sole traders seem to achieve a better bottom line. When it comes to the growth in assets, being the other form of financial outcome, sole traders achieved 10.50%, partnerships 12.69%, but companies 1.98%. This

Table 8.3. Characteristics of the managers and businesses in each ownership category (average statistics for all managers/businesses in each column rounded to one decimal point). (From Nuthall and Old, 2017.)

	Ownership form (no. of businesses; ave. business size (production units))			
Variable	Sole trader (230; 506)	Partnership (458; 536)	Company (60; 1530)	*F*-statistic significance probability
Age (years)	54	53	48	0.000
Formal education[a]	3.2	3.2	3.6	0.079
Ave. grade last year of study (%)	67.3	61.5	65.0	0.063
Gender (1 = 'female', 0 = 'male')	0.15	0.11	0.12	0.000
Number of children	2.2	2.5	1.7	0.002
Years in current business	30.0	29.4	25.5	0.045
Number of labour units incl. manager	2.2	2.8	3.9	0.002
Number of separate businesses	1.6	1.8	2.0	0.005
Management style variables[b]				
Anxiety	3.4	3.3	3.1	0.107
Conscientiousness	2.8	2.7	2.5	0.064
Agreeableness	3.0	3.1	2.7	0.064
Openness	2.1	2.1	1.9	0.083
Extraversion	2.8	2.6	2.5	0.090
Objective variables[b]				
Lifestyle and leisure	1.9	1.9	1.7	0.121
Enjoyable work	2.2	2.1	2.1	0.007
Monetary return and growth	2.8	2.8	2.5	0.104
Important to reduce risk	2.2	2.3	2.5	0.383
Profit per labour unit (NZ$)	46,280	50,420	57,280	0.612
Years in business squared	901.8	862.0	649.2	0.047
Net assets per labour unit (NZ$)	2,106,000	2,206,000	2,722,000	0.207
Entity assets – shares/trusts (% of total)	41.0	44.2	53.6	0.285
Spent on advice (NZ$)	2690	2530	5400	0.090

Continued

Table 8.3. Continued

Variable	Ownership form (no. of businesses; ave. business size (production units)) Sole trader (230; 506)	Partnership (458; 536)	Company (60; 1530)	*F*-statistic significance probability
Hours of advice on governance and succession	6.4	5.9	8.2	0.827
Involve others in risk advice (1 to 5 scale)[c]	2.8	2.6	2.1	0.003
Aware of governance structures availability (1 to 5 scale)[d]	2.4	2.4	2.4	0.959
Percentage of decisions made by manager	88.1	84.5	90.4	0.081
Management ability (%)	60.3	61.4	60.9	0.322
Know finances in detail (0 to 4 scale)[e]	2.2	2.5	2.7	0.009

[a]Scale: 1 = 'primary', 2 = '< 4 years secondary', 3 = '> 3 years secondary', 4 = '< 3 years tertiary', 5 = '> 2 years tertiary'.
[b]Personality style and objectives rated on a 1 to 5 scale with 1 = 'complies with description', 5 = 'does not comply with description'. These 'variables' are summarized by taking an average of the original variables classified according to which personality style factor they determine. And similarly for the objectives.
[c]Scale: 1 = 'completely' to 5 = 'not at all'.
[d]Scale: 1 = 'fully aware' to 5 = 'not at all aware'.
[e]Scale from 0 to 4 depending on the knowledge of business's financial performance (0 = 'no knowledge', 4 = 'full knowledge').

last figure seems well out of line, but perhaps the companies were carried away with their land purchase decisions over investing?

When the managers were asked for a happiness rating on their existing ownership structure, they were largely content with little interest in changing. Perhaps all these positive responses to being a sole trader relate to the intimate relationship between manager and business that exists and its impact on production efficiency success.

In contrast, as most primary producing ventures experience considerable production and profit variations year upon year with the vagaries of weather, disease and international markets, corporate ownership systems are less able to cope with ups and downs. Business families are often able to reduce the cash take and better cope with the variability. Corporate owners can compare rural outcomes to urban investments which are largely relatively stable and likely conclude non-rural investments have advantages. The exception is the investor who has a real interest in things rural and may well enjoy visiting their investment.

It is also interesting to note an increase in pluriactivity as business families seek greater stability and diversification. Family companies can follow suit, but public companies with their paid managers are likely to have less interest in these, frequently 24 hour, activities such as tourism.

The logistic regressions referred to explained around 60% of the variance. It is interesting to ponder the other 40%. The information held did not cover issues like the family history associated with the business, nor succession factors. Perhaps these have an influence on the strength of the associations, which will have to await further research. However, this does lead on to the next section covering the impact of personological variables on succession issues.

Business Succession

Introduction

Social and economic health in SME businesses, and their impact on the rest of the community, requires the successful succession of businesses from one generation to the next. In that large numbers of SMEs are family-based, the issue becomes one of family succession systems. The problem is that managers are particularly bad at thinking ahead to these issues, and relatively backward in taking action. This section contains an outline of research (Nuthall and Old, 2016) that explored this issue leading to ideas on what might be required to reduce the impediments to succession. One of the hypotheses was that a manager's personality (management style) was a major factor in this commonly experienced 'laggardly' behaviour.

Nuthall and Old (2014) found managers aged over 65 years had only started transferring assets 10.4 years earlier in a jurisdiction that, at the time, required paying a tax on gifting. Despite this, it is interesting to note the tax encouraged starting transfer gifting as early as possible, allowing being below the gift tax threshold. And for managers aged between 56 and 65 years the figure was 5.9 years. It was also revealing that 68.3% of the managers in total had not transferred any assets at all to the next generation. Similar situations exist in most countries. In North Carolina it was found only 30% of managers who had reached 55 years had identified a successor, though in the UK the equivalent figure was 50% (Lobley *et al.*, 2012).

No doubt there are many reasons why managers are reluctant to plan succession, and to act. Managers probably relate retirement to mortality, and it is certain managers become passionate about their life work in developing an SME business. Thoughts going through their minds over how they would obtain satisfaction if forced away from their nirvana no doubt abound. Retirement likely holds visions of being totally misplaced. While these are all understandable reactions, the data in general shows the managers who plan early for succession leave the business and new manager in a better situation:

- profit wise,
- asset wise,
- efficiency wise, and, most importantly,
- family relations wise;
- allowing successful arrangements for the retirees, and
- promoting fair distribution of assets to all heirs.

Furthermore, evidence is available that poor succession planning tends to lead the business into another family's hands; for example, in the USA it is infrequent for businesses to survive more than two generations (Le Breton-Miller *et al.*, 2004).

To explore these succession issues a model was developed to quantify the relationships using a set of manager data obtained from a survey. The sections that follow explain the model, the collection of the data and the results from the analyses.

Model

The manager data available contained both the manager's plans over asset distribution as well as the level of assets already distributed. This allowed exploring the differences between the potential and actual distribution as well as the reasons for the differences.

Given this situation the obvious core model to use was the Theory of Planned Behaviour (TPB) (Ajzen and Fishbein, 1980) in which it is hypothesized that a decision maker's action is dependent on three factors. In this case they cover the manager's intentions (attitude (ATT)), what the manager perceives as society's views of the issues (social norms (SN)) and what the manager sees as restrictions on what is possible due to technical and resource factors (perceived behavioural control (PBC)). Each factor is made up of a number of variables, as explained later, which all lead to the

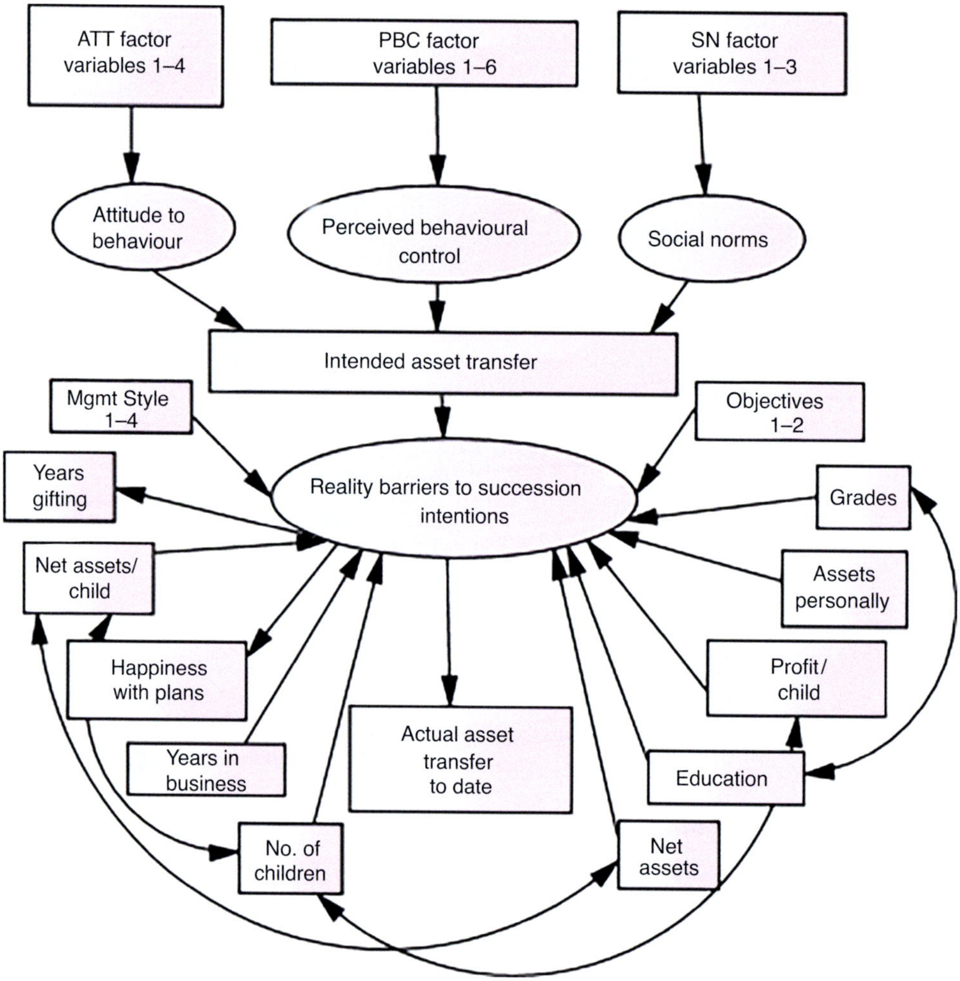

Fig. 8.3. Structural Equation Model of succession intentions and asset transfer level. (From Nuthall and Old, 2016.)

manager's intention over succession. But in reality, intention seldom leads to completely compliant action.

This difference between intention and actuality is caused by various barriers. This is all represented by the sections in Fig. 8.3. The ellipses at the top of Fig. 8.3 represent ATT, PBC and SN, with arrows leading into the ellipses representing the variables influencing each. In the middle is a rectangle representing the intended asset transfer. This then leads to the bottom rectangle giving the actual 'to date' transfer via an ellipse representing the barriers to the manager's succession intentions, thus explaining the difference between planned and actual.

Various factors are hypothesized to influence the barriers as shown by a range of rectangles from which arrows lead into the 'barriers' ellipse. Each rectangle represents a number of sub-variables. The objective of the model is to find out the importance of the factors which impinge on the barriers to discover where attention can be focused to improve succession arrangements.

Logic, the literature and experience suggest the variables to include in the model. The rectangles in the model represent each class. Generally any variable in a rectangle is an observed variable whereas variables in ellipses are what are called 'latent' constructs. These variables, as noted earlier, are not directly observed but are implied through the variables' originating arrows which represent the direction of influence.

This overall model (Fig. 8.3) is a mixture of a TPB core together with the concept of the reality barrier and the variables creating the barriers. The top half of the model reflects the TPB, and the lower half the barrier system and its determinants. To solve this total construct, in which the directions of influence lines represent linear regressions, a simultaneous solving system is required. Conventionally the total system is referred to as a Structural Equation Model, as described earlier in this book (Chapters 3 and 7), which is solved using structural equation modelling (SEM), an iterative maximum likelihood method (Bowen and Guo, 2012).

Data and analyses

It is likely that the variables to include cover family issues such as family size, parent's age, parent's health, education, children's ages, parent's experience and managerial ability, and similar information on possible successors. If it could be measured, family relationships are likely to be a factor given research has shown family cohesion can have a major influence on agreements resulting from family conferences either with, or without, third-party assistance. Tax and laws relating to succession can also be important influences, but across businesses are identical in any one jurisdiction preventing meaningful, or relevant, comparisons.

Also relevant, as has been stressed and included in the model, is the set of objectives held by the manager as they will influence succession intentions. Personality issues (management style), such as conscientiousness and agreeableness, are also important through their impact on the determination to achieve objectives and a willingness to reach a whole-family agreement. In most contemporary societies equality between heirs is usually important. Furthermore, succession, as has been noted, can also be impacted by the business ownership structure.

The data for the calculation of the model parameters was obtained through a postal survey of 2300 randomly selected stratified managers as explained earlier in the chapter (Nuthall and Old, 2014). The large number of variables in each category was factorized to give such constructs as the management style and objective variables (as explained previously). In addition, the variables giving rise to the 'experience', ATT, PBC and SN were similarly factorized to give five experience variables (the names reflect the more important sub-variables in each factor):

- 'years of experience',
- 'people centric experience',
- 'decision allocator',
- 'major problems experienced', and
- 'mistake learner';

four ATT variables:

- 'age imperative',
- 'family empire',
- 'challenge acceptor', and
- 'expander with professional help';

six PBC variables:

- 'mixed cash/enjoyment seeker',
- 'conservative risk remover',
- 'child provider',
- 'efficient technologist',
- 'doubtful changer', and
- 'asset controller';

and three SN variables:

- 'community satisficer',
- 'colleague consulter', and
- 'professional consulter'.

All other variables were taken directly from the completed questionnaires. Examples include education levels, number of children and their ages, asset valuations and similar.

Table 8.4 contains the more significant variables from the SEM analysis. The coefficients presented are the standardized values which allow comparing each variable despite the variety of measurement units, together with their significance probability levels. Each line gives, effectively, the linear regression coefficient for each arrow in Fig. 8.3. The first named

Table 8.4. Coefficients from the SEM model. Regression values for each arrow as well as their standardized value and significance probability. Variables with significance probabilities over 0.28 not included. (From Nuthall and Old, 2016.)

Variable pair (each pair represents a line in the model; the arrow indicates the direction of influence)	Standardized coefficient	Significance probability
Age imperative → attitude	1.91	0.056
Expander with professional help → attitude	−1.35	0.178
Community satisficer → subjective norm	1.25	0.212
Colleague consulter → subjective norm	−2.12	0.029
Professional consulter → subjective norm	−1.37	0.171
Mixed cash/enjoyment seeker → perceived behaviour control	−1.21	0.227
Conservative risk remover → perceived behaviour control	3.19	0.001
Child provider → perceived behaviour control	2.16	0.031
Efficient technologist → perceived behaviour control	−2.07	0.039
Consultative logician family and friends (management style) → barriers	−2.15	0.032
Way of life (objective) → barriers	1.08	0.281
Family support (objective) → barriers	−2.02	0.043
Grade last year of study (%) → barriers	1.22	0.223
Assets held personally → barriers	−3.74	0.000
Net assets → barriers	1.24	0.214
Number of children → barriers	−1.09	0.274
Profit per child → barriers	−2.29	0.022
Years of experience on business → barriers	2.43	0.015
Barriers → sum actually given to date (NZ$100 units)	12.74	0.000

variable is the originating rectangle and the second the target variable. The last ten lines, other than the very last, give the variables giving rise to the 'barriers' latent variable. The last line reflects the impact of the barriers in total on the assets actually transferred – thus the highly significant 12.74 coefficient shows just how influential the barrier is relative to the size of all the other coefficients.

In terms of the variables influencing the asset transfer objective, the 'conservative risk remover' factor in the PBC is the most influential as might be expected, and the 'child provider' component also important, again as would be expected. At the other end of the spectrum the 'colleague consulter' is influential through the SN reflecting a manager taking notice of what others think.

For the influences on the barriers, the 'years of experience' and the level of personal assets have important impacts, as does the profit per child reflecting the relationship between the number of children and profit. But what is really interesting is the lowering of the barrier with a management style involving consulting family and friends – this reflects the often-mentioned need for good family relationships. Similarly the 'family support' objective is important.

An important benefit from the modelling is the ability to calculate the barrier coefficient for each manager using the coefficients of the variables giving rise to the barrier value. Using the value for each manager it was possible to score each with their percentage barrier level (a barrier index) after obtaining the average and range values. The distribution of the index approached normality and had a mean of 47.4% and a standard deviation of 19.5%.

Using this percentage the managers were divided into three groups: the managers with a score greater than 69%, those with 69 to 31%, and the rest with a score less than 31%. A comparison of the top and bottom groups' characteristics provides information for determining the managers likely to be 'laggards' relative to those who are logically moving ahead on succession issues. Table 8.5 contains this comparison together with significance probabilities of the differences.

There are very clear differences between the groups, with the 'low barrier' managers having passed on 77% of assets relative to the 16% for the 'high barrier' managers. And also of note is the profit level per child being much lower for the 'low barrier' managers despite family size being much the same. Other differences are the higher education level of 'low barrier' managers and their higher grades, which is the more significant of the two.

The major interest in the differences is indeed the comparison of the management style and objectives of the managers. The significant differences in management style involve the types 'correctness seeker', 'consultative logician friends and family' and the 'thoughtful creator'. Logically it can be understood why these characteristics would lead to differences in the barrier levels.

For the objectives, the significantly different 'way of life' objective increases the barrier level which most likely means managers who really value living and working on 'their' property are reluctant to move on to a different phase of their life. In total contrast, managers who are 'reluctant managers' are keen to move on, exhibiting a low barrier score.

Messages

Overall, once again, it is clear the personal characteristics of the people involved certainly influence attitudes to succession and retirement.

In this sense, the managers were asked how happy they were with their succession plans and progress. These data were used in a regression analysis to explain succession happiness which had, albeit significant, a 40% explanation of the variance, showing that additional factors than were available featured in their happiness levels. These would explain, hopefully, the other 60%.

However, it was clear happiness decreased with the level of assets planned to be passed on, reinforcing that many managers reluctantly move into retirement. Many, however, are forced into this situation through ageing bodies and poorer health. And as might be expected, several management style and objective factors appeared in the equation including the benign trait, 'community satisficers' and business expanders. Interestingly, managers who had experienced major problems over their business life tended to be happier with their succession plans.

To reduce the barriers, working on objectives and management style will promote action – which is one of the clear messages. But there are other strategies that can help which are often highlighted in the generally available assistance pamphlets and books. Ideas such as bringing in a trained mediator, ensuring family communication occurs successfully and regularly involving all participants, starting early, allowing flexibility (as conditions do change) and bringing in professionals (e.g. lawyer, banker, accountant etc.) early on, all abound.

Manager Anxiety

Introduction

Chapter 6 contained a discussion on manager stress indicating just how common it is among managers and other business people. It also outlined the methods of coping with the problem. This chapter extends the discussion through covering an associated research project.

Another term for stress is 'anxiety', with the two words being synonymous as far as this discussion is concerned. Anxiety is counterproductive in many ways and needs to be addressed if at all possible. In its extreme form it can totally

Table 8.5. A comparison of variable means for managers exhibiting a 'barrier index' percentage greater than 69% and less than 31%. Variable means and significance probability of the differences (*t*-test probability). (From Nuthall and Old, 2016.)

Variable	Mean for high barrier group	Mean for low barrier group	Significance probability of difference
style[a] – consultative logician community	−0.068	−0.318	0.391
style[a] – correctness seeker	−0.033	0.655	0.009
style[a] – consultative logician family and friends	0.347	−0.892	0.000
style[a] – conscientious planner	−0.177	−0.062	0.663
style[a] – thoughtful creator	−0.090	0.341	0.123
style[a] – benign manager	0.055	0.264	0.432
Objective[a] – balanced	−0.078	−0.322	0.362
Objective[a] – risk remover	−0.137	−0.189	0.863
Objective[a] – way of life	−0.031	0.595	0.036
Objective[a] – reluctant manager	0.350	−0.721	0.000
Objective[a] – community supporter	0.027	0.301	0.342
Objective[a] – family supporter	−0.050	−0.110	0.833
Manager age (scored 1 to 6)	4.87	5.05	0.506
Gender (0 = 'male', 1 = 'female')	0.053	0.048	0.935
Business type (scale 1 to 11)	4.2	3.2	0.196
Years in current business	22.97	41.29	0.000
Percentage of decisions made by manager	84.1	74.0	0.154
Education level (scale 1 to 5)	3.4	3.7	0.285
Grade last year of study (%)	61.0	70.5	0.011
Net assets (NZ$)	5,300,125	8,695,238	0.131
Assets held personally (% of all)	82.32	5.38	0.000
Number of children	2.82	2.67	0.628
Profit per child	106,423	48,758	0.343
Dollars promised in succession	1,696,600	2,293,400	0.085
Dollars actually given to date	279,000	1,762,800	0.000
Percentage actually given to date	16.44	76.86	N/A[b]

[a]Factor score for the management style factors OR objective factors, respectively. See Chapter 7 for definitions of the management style and objective factors.
[b]Not applicable as calculated.

disrupt lives as shown by the not uncommon occurrence of manager suicides. Managers with extreme cases of anxiety, and depression, should immediately seek professional help.

The trouble is most managers are particularly bad at talking about their stress and anxiety, and even worse at seeking help. Every effort must be made to overcome this reluctance with business family and friends encouraging openness and a search for assistance. Managers themselves must clearly play their part and encourage each other to discuss issues and seek help. In many areas it is now possible to access manager groups and organizations set up specifically to help people who are experiencing anxiety or depression.

Most managers will have experienced anxiety at some stage, and some experience it on a constant basis. While some anxiety can encourage attention to management detail, and can be useful, in general anxiety is to be avoided using the various approaches available. Anxiety is very counterproductive with its impact on:

- biased decision making, and thus profit and other achievements; and
- the reduction of enjoyment together with unpleasant worry.

By its very nature, SME business management is stressful with anxiety occurring in a wide range of situations. These are listed in Chapter 6. Also listed are a range of coping mechanisms. However, in many cases a better way to overcome anxiety is to modify a manager's core characteristics that give rise to the anxiety. Such an approach is more likely to be long lasting with the resultant mental reactions more positive.

This section reports on a research project designed to explore the basic human factors associated with anxiety in order to clarify the characteristics that should be addressed if at all possible.

In a general sense, besides using the coping mechanisms outlined in Chapter 6, two approaches to anxiety are:

- removal of the factors giving rise to anxiety (e.g. using diversification to reduce income variability, using contractors instead of employing help if the latter creates anxiety etc.); and
- changing a manager's mental reaction to anxiety-creating situations.

The research reported here (Greig *et al.*, 2020) explores the issues relating to the second method. However, any manager must consider both methods, possibly with professional help in deciding the mix.

This section proceeds by further exploring anxiety issues as well as suggesting factors that are at the root of anxiety through explaining a model representing the issues and logic of the variables included. It also covers the collection and analysis of anxiety data from a large group of managers, and, finally, using the analysis, proffers some conclusions over reducing anxiety levels.

Anxiety modelling

Not unexpectedly, Greig *et al.* (2020) hypothesize that anxiety is a function of human characteristics together with the features of each particular business and business family. Clearly, some business environments are more difficult to manage, and some have much more variability than others. And just by chance, a manager might be faced with difficult business colleagues and employees. This all adds to anxiety for some people.

The situation is summarized in Fig. 8.4. There are many sources of anxiety on a business including weather variations, disease concerns, banking difficulties, variable markets and import systems, and changing rules and regulations. The list goes on. However, many can be summarized, as shown in the central ellipses in Fig. 8.4, down to:

- business working environment anxiety;
- family-based anxiety;
- rules/regulations anxiety; and
- financial issues anxiety.

These can be further combined into 'business issues' and 'financial issues' anxiety as shown in the lower ellipses.

Then, in the lower rectangles of Fig. 8.4, extreme anxiety, being significantly different from normal run-of-the-mill anxiety, is highlighted for the:

- business issues; and
- debt area.

Extreme anxiety was defined as 'quite significant anxiety' in the survey questionnaire.

So far in Fig. 8.4 it is only the various anxieties that have been highlighted (the manager questionnaire asked for ratings on 'production issues', 'product prices', 'financing/debt', 'mismanagement', 'family issues', 'employees', 'rules and regulations', 'environmental regulations', 'health', 'work conditions', 'extreme weather', 'isolation' and 'time pressures'). But what of the variables giving rise to anxiety? The rectangles at the top of Fig. 8.4 reflect these in summary form. Many of these are listed in Table 8.6 presented below. But suffice to note they cover variables such as the manager's age and education, the manager's spouse's attitudes to debt, self-assessed forecasting skills, details of the children, income levels, and many more.

In addition it will be noted the manager's management style and objectives are highlighted as would be expected because it is these variables that are hypothesized to be important, particularly the anxiety factor in management style (personality). In addition, again as would be expected, the manager's locus of control features in its own rectangle.

To assess the importance of the model components, a postal survey was carried out on a stratified (business type, region and size), randomly selected sample of businesses resulting in over 400 manager replies. The questionnaire

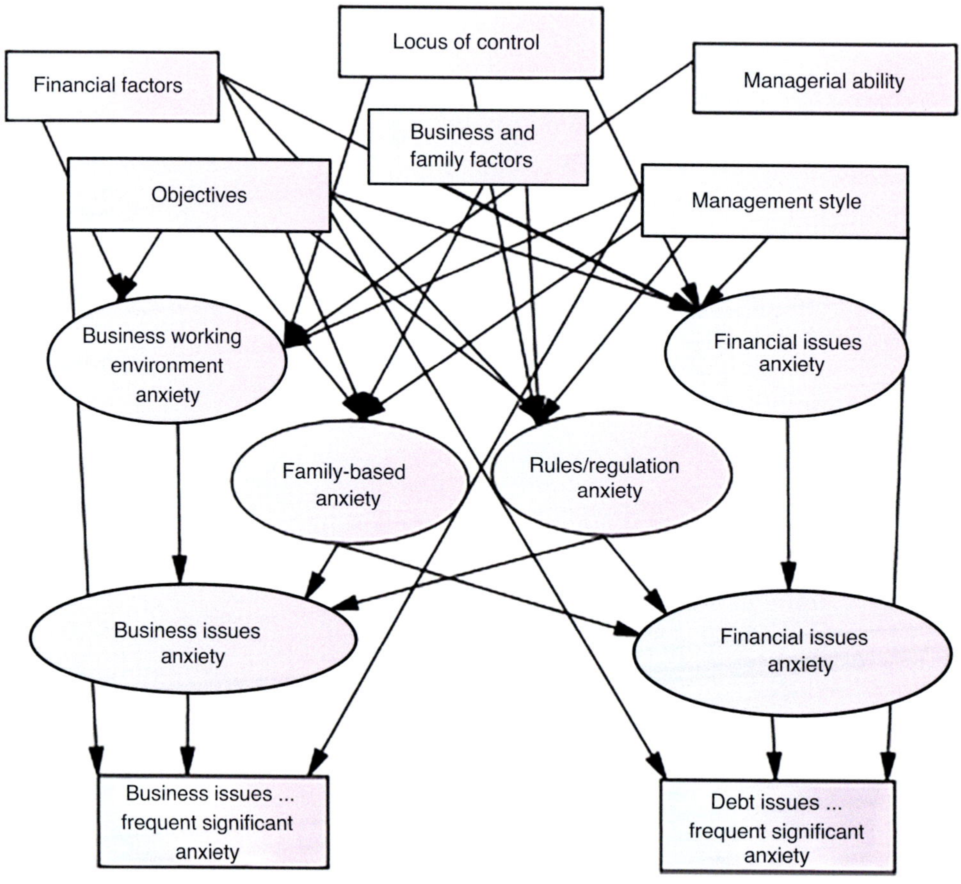

Fig. 8.4. A schematic representing the manager anxiety problem and the associated variables. (Modified from Greig *et al*., 2020.)

Table 8.6. A comparison of variables for managers experiencing high, as against low, frequency of significant anxiety for (i) production/control issues and (ii) debt issues.

	Production/control issues			Debt issues		
Variable	High anxiety (mean)	Low anxiety (mean)	Significant difference probability (*t* test)	High anxiety (mean)	Low anxiety (mean)	Significant difference probability (*t* test)
Age (scale 1 to 6)	4.88	5.38	0.000	4.99	5.07	0.552
Education (scale 1 to 5)	3.32	3.24	0.668	3.27	3.27	0.997
Grade (%)	61.97	60.74	0.521	59.99	60.73	0.514
Gender (1 = 'female', 0 = 'male')	1.93	1.91	0.608	1.96	1.90	0.126
Style – extraversion[a]	–0.020	0.028	0.777	0.057	0.014	0.765
Style – conscientious[a]	–0.087	0.122	0.180	0.084	–0.038	0.392
Style – anxiety[a]	–0.367	0.569	0.000	–0.476	0.382	0.000
Style – open[a]	–0.040	0.059	0.527	–0.170	–0.017	0.284
Style – agreeable[a]	0.115	–0.231	0.044	0.149	–0.126	0.057
Objective – work[a]	–0.033	–0.124	0.577	0.158	0.200	0.012
Objective – profit[a]	–0.225	0.318	0.001	–0.299	0.167	0.001
Objective – leisure[a]	0.129	0.045	0.578	0.029	–0.069	0.496
Objective – risk[a]	–0.053	–0.055	0.992	–0.166	–0.083	0.555
Objective – community[a]	0.249	–0.082	0.044	0.188	–0.012	0.179
Locus of control (LOC) – uncontrollable[a]	–0.265	0.448	0.000	–0.190	0.306	0.000
LOC – conservative[a]	0.098	–0.052	0.335	0.046	–0.016	0.670
LOC – determined[a]	–0.118	–0.057	0.708	0.047	0.019	0.591
LOC – planner[a]	–0.058	0.153	0.197	0.004	0.019	0.921
LOC – people[a]	0.142	–0.057	0.234	0.178	–0.046	0.141
Business intensity[b]	3.46	3.44	0.933	3.68	3.33	0.098
Productive efficiency[c]	0.943	0.991	0.401	1.00	1.00	0.988
Gross income/ labour unit (NZ$)	297,160	254,780	0.166	295,320	255,120	0.695
Personal ownership %	91.96	92.87	0.479	91.72	92.95	0.804
Financial knowledge[b]	1.93	2.13	0.222	2.15	2.05	0.698
Forecasting ability[a]	–0.109	0.066	0.169	–0.006	0.007	0.914
Managerial ability[b]	2.96	3.36	0.045	1.26	1.25	0.694

Continued

Table 8.6. Continued

Variable	Production/control issues High anxiety (mean)	Production/control issues Low anxiety (mean)	Production/control issues Significant difference probability (*t* test)	Debt issues High anxiety (mean)	Debt issues Low anxiety (mean)	Debt issues Significant difference probability (*t* test)
Spouse debt attitude[b]	2.68	1.99	0.001	2.92	2.04	0.025
Spouse debt influence[b]	2.31	1.96	0.120	2.52	1.82	0.001
Procrastination[b]	3.45	3.94	0.008	3.35	3.83	0.004
Number of children	1.93	1.37	0.017	2.01	1.60	0.458

[a]Factor scores as explained earlier in this chapter and in Chapter 7.
[b]Score from 1 to 6.
[c]Relative to top managers.

was eight pages covering manager objectives, management style and locus of control, all using the sets of questions described in previous chapters. Similarly, information on the manager and business family was obtained as well as a whole range of anxiety ratings and output quantities. A copy of the questionnaire is given in Greig *et al.* (2017) (available online). Anxiety levels were assessed on a ten-point scale with 1 = 'little anxiety' through to 10 = 'great anxiety'. The information was factorized into summarizing variables for the areas in which many questions were used (such as the management style set with its 25 statements to be rated).

The information was also fed into a number of regression equations and comparisons as reported in the next section.

Results of the analyses

The businesses in the survey covered a range of types. For comparative purposes a business intensity index was created, with one representing larger extensive-type systems through to six for smaller intensive operations. Financial knowledge was also introduced through a self-rated score from 1 to 5 as it was thought this would impact on feelings of anxiety. Similarly, the idea that a procrastination tendency might well impact on anxiety, in addition to personality issues, was catered for with a self-rated five-point scale using the statement 'I tend to wait too long before deciding and acting'. Various other similar ratings were also covered as will be noted in the result presentation below.

In the first instance, to assess the impacts of anxiety on outcomes, regression equations were calculated to explain the business's productive efficiency, the level of equity (and thus debt), return on capital and the average change in cash return over the last 5 years (to even out the common yearly fluctuations). The four equations explained, respectively, 34, 30, 45 and 15% of these outcome measures, and each was a highly significant equation. Respectively, the anxiety variables accounted for 50, 86, 46 and 55% of the explanations. The point is largely proven.

To explain the core factors in a manager's anxiety levels, seven further linear regression equations provided food for thought. The first three covered explaining manager levels of total anxiety (combined business, financial, rules/regulation and family), frequency of significant business-based anxiety and frequency of significant debt-based anxiety. These equations explained by far the majority of all anxiety with, respectively, 99, 73 and 97% of the variance covered, all at highly significant levels.

Of the human characteristics, starting with management style, the style components explained 19, 17 and 17% of the different anxieties, respectively. For the objectives, the equivalent figures were 24, 9 and 14%, and for the locus of control the degree of anxiety explained was 33, 22 and 11%. While each set of human characteristic variables contributes to anxiety, as might be expected, a manager's locus of control stands out as an important factor in frequent and significant anxiety.

Breaking anxiety down into business-based, financial, rules and regulations, and family sourced, the equations explained 58, 82, 81 and 59% of the variance, respectively. And for the human characteristics, management style accounted for 50, 25, 21 and 35% of the total explanation of the various anxieties, respectively. For the manager objectives, the corresponding figures were 14, 29, 29 and 12% for each type. There are clear differences between these two general factors in their influences. And the same applies for the locus of control with the proportions of explanation being 15, 31, 57 and 26%, respectively. Overall, as would be expected, it is clear the human factors are very important in explaining anxiety. But it was important to verify this with the analyses.

If the percentages for each variable grouping are added to give the percentage of human factors influence for each general type of anxiety (business, financial, rules and regulations, and family-based), the totals come to 79, 85, 57 and 73%, respectively. The highest is for financial anxiety, with only 15% being attributed to variables other than those featured above such as age and procrastination. The lowest total is for 'rules and regulations' anxiety, but still a majority figure at 57%.

The clearest way to demonstrate the contribution of personal characteristics to anxiety is to group the managers into classes based on their degree of anxiety and to compare each group's mean characteristic levels. Table 8.6 contains the results for the business anxiety and debt anxiety classifications.

As would be expected, the personality factor 'anxiety' is totally different for the 'high anxiety' compared to the 'low anxiety' manager group, and the large differences are highly significant as shown by the *t*-test probability values. Other than factors like the average age for each group being significantly different ('low anxiety' managers are older), and differences in managerial ability, procrastination, number of children, and spouse's attitude and influence on debt, agreeableness levels are different. Furthermore, profit motivation (objective), community attitude and the locus of control factor 'uncontrollable' are all similarly significantly different, further expressing the importance of human characteristics in anxiety levels.

Messages

If a manager wishes to work on reducing anxiety to gain the various benefits outlined, the results give clear pathways to work on. For modifying a manager's management style it is likely to be necessary to enlist professional help of various sorts, such as a counsellor and/or consultant, but also to form discussion groups of like-minded managers to provide ideas and support, and also to mutually critique, change efforts. 'Significant others' might also provide a sounding board for the change efforts. An important issue is recognizing where change is most likely to be beneficial. This requires strong personal critiquing skills, and acceptance of others' suggestions.

For any group or professional working with managers it is useful to predict which managers might benefit from support and assistance. This is where a knowledge of the important anxiety-creating human characteristics come into play, as do the management style and the locus of control tests in addition to the assessment of a manager's objectives. The list of objectives can lead to a discussion on how appropriate they are, remembering in the end it is the manager's, and business family's, prerogative to set their own goals and values.

Concluding Comments Overall

The studies presented could be further extended in this same vein, but those covered suffice in strongly suggesting that the managers' human factor, or personal characteristics, play a large part in the success of most aspects of SME business operations.

The examples cover disparate issues ranging from succession to expansion. Earlier chapters focused on the importance of the human factor in, largely, managerial ability and successful business operations, as do subsequent chapters. This chapter, while clearly still relevant to managerial ability, explores issues that have both long-lasting impacts on manager efficiency as well as impacts on manager satisfaction and the achievement of family objectives.

Another important factor so far only touched on, which involves the majority of businesses, is debt and equity. A recent study of these aspects shows how manager characteristics strongly feature in the propensity to reduce debt (Greig *et al.*, 2019). The list goes on.

What is also interesting is how in each case the components of management style, objectives and the locus of control that are important for each studied problem tend to change. Logic would suggest this because the nature of each situation, and its influence on the life of a business, differ. It is also certain a business encompasses a very wide range of issues, right from the psyche of the people involved through to the weather impacts of sun spots. The topics not influenced in some way or other by a manager's characteristics are few and far between. This situation creates many challenges for a manager as each situation interacts with their personalities, objectives and locus of control.

These wide-ranging impacts also provide challenges for professionals involved in extension, consultancy and policy. Each must isolate the human factor impacts and fully allow for them in their work. In addition, where managers wish to make permanent, and improving, changes to their psyche further professionals are likely to become involved.

The effects of a manager's human factors are all pervasive.

9 More on Objectives: Family Influences, Origins and Modification

Abstract

Businesses' objectives should direct decisions in all situations and provide the yardsticks for assessing success. This chapter looks at the factors influencing the origins of a manager's objectives. In many cases the families associated with a business may exert an influence, as will 'entrepreneurship'. Families are often in conflict with the business over resources, cash and time. Good family relations are critical to resolving issues involving needs, desires, attitudes to control, and future directions. However, dysfunctional families do exist where help could be needed. The influence of a spouse can be significant and is discussed. Marital harmony is also important with personality impacting on good relationships. Values (instrumental, social, expressive, intrinsic) give objectives which lead to goals. Objectives change through the life cycle and emotions impact. A statistical analysis is presented to give clarity to objectives. Age, gender, conscientiousness and other personality traits are examples of significant variables. Another analysis quantifies Locus of Control and its influence, and entrepreneurship is discussed. Personality provides causation in both cases. Finally, the case managers provide interesting examples.

Introduction

A manager's objectives strongly impact on the decisions made. This is one of the reasons why the outcomes from every SME business tend to be different, as each manager's objectives will be unique. Also important to decisions in most SME businesses is the manager's family including a spouse. Thus, objectives and families are further considered in this chapter.

Part of comprehending a manager is the understanding of his or her objectives, and the origins of these objectives. So, one of the first steps in helping a manager is determining whether his/her objectives are correctly stated. Progress cannot be measured without these yardsticks. But, while determining the objectives (perhaps using the questionnaire listed in Chapter 3, or through careful observation) is important, of even greater value is the understanding why the manager holds the particular set. Possibly the manager has concluded incorrectly and so discussion and assessment may lead to modifications. To this end one of the sections in this chapter contains a discussion on objectives and the influence of the family. Similarly, as a manager's 'locus of control' (LOC) may be important in constricting progress, factors which give rise to a particular attitude are considered with a view to understanding a manager's LOC, and what might be done about improving the situation.

Everywhere in the world most SME business tends to be both owned and managed by a family or similar entities. Thus, an important influence on the business operation is this family, both spouse and children, and in some cases, even the earlier generation. Accordingly, a discussion on their influence is included in this chapter. The family is part and parcel of the human factors.

When talking about objectives, the question of entrepreneurship should also be considered. In an earlier chapter, motivation was referred to with the comment that it is encompassed by managerial style rather than being an independent factor. The same might be said of entrepreneurship, but as it is a much-discussed subject, it will be considered further.

A manager who is at the forefront of innovations and always seeking new products, methods and 'off-business' ventures that might perhaps build on his business success is often referred to as an entrepreneur. Such managers hope to better achieve their objectives by being adventurous and creative in their business operations. A stock manager might, for example,

DOI: 10.1079/9781800627246.0009

buy a chain of specialist butcher shops to market his, and his colleagues', produce in the hope of making better margins, and increase his wealth and power. What are the features of such people, and can others acquire these features where they aspire to expand?

The Family Influence

Family relationships are critical to the success of a business. SME business is largely unique in the business world in that the family is often frequently and intimately involved through both living on or near the business, and probably being part owners. Very few urban businesses face this situation, which is both an advantage, and, in a few cases, a hindrance. Managers and their families often live in quite isolated situations relative to urban situations, and for many days a manager and his spouse will have no contact with other people, and similarly for any family member working in the business. This is an important reason why the family must be in harmony, and, if not, achievement of their objectives will be made difficult. And while corporate ownership of the business is increasing in many parts of the developed world, it is still in the minority as many so-called 'company rural businesses' are in fact family-owned and -operated. For example, in Colorado, around 90% of the SME farms are family-owned and -operated. And in New Zealand the ownership is held by:

- the manager in 30.5% of the cases;
- family partnerships 57.5% of all cases; and
- family corporations, trusts and public company business for the remaining 12%.

Where there are truly corporate situations with the owners totally divorced from the management, the family influence is largely non-existent, with the manager simply carrying out the board's instructions on the objectives and the goals. However, the majority of SME business is not in this position, and it is doubtful whether this will ever be the case, as most business families are prepared to accept a lower return than that required by corporate investors. Furthermore, SME rural business is often a 24 hour, 7-days-a-week operation and there are few employees happy to work such long hours without an intimate involvement in sharing the benefits and costs. For this reason, sharing some operational arrangements can be successful as the manager has a direct interest in, say, checking out a burglar alarm in the middle of the night, or perhaps shifting gear in the middle of the night when a flood is imminent. To achieve success, living on or near the business is often a prerequisite. Truly, the human factor is crucial in the life of a business.

With family business there is competition for the:

- resources;
- cash; and
- time.

How much of the available time is devoted to production, and how much to the family and leisure? When the products are sold, what is the allocation of the proceeds to costs, business improvements or maintenance, and what percentage ends up on consumption and the family home? Clearly there can be considerable conflict between a spouse and the manager, and the children, not to mention other family members who might be involved such as a father or working children. Family relationships and procedures are clearly important to ensure harmony and good decision making for both the business and the family (Fig. 9.1).

In by far the majority of cases, a spouse will be a wife, and at some stage in the life of a business, a mother. While there is a trend for females to be business managers and owners in their own right, the percentage is still very small. But that does not mean the spouse is not involved in both business work and decision making. This participation also means a spouse is increasingly involved in setting objectives, though often this is not directly expressed in setting out a list, but rather implied in the decisions made, such as devoting money to upgrading the business family's house or on education or perhaps leisure activities. The survey data suggests that spouse involvement in a range of decisions covers the full spectrum from complete involvement through to none at all.

While many families work harmoniously over objectives, decisions and relationships, it

Fig. 9.1. Competition for resources between the business and the household can sometimes be intense and problematic.

is certainly not always the case. Harmonious relationships depend a lot on:

- the personality of the family members;
- the interests of the members of the family group; and
- how well they listen to and consider each other.

The success of the business operation does influence the relationship and if all outcomes are considered at least satisfactory, there are less likely to be tension spots. But where outcomes are well away from what might be expected, this can erupt in poor relationships and a dysfunctional situation that only outside help, or valuable leadership from one of the family, can solve and lead to better outcomes. Thus, an observer must be careful to assess not only the manager, but the whole family unit and how it cooperates and works positively.

Harmonious families will have:

- good communication skills; and
- open relationships.

They will be constantly talking to each other about their wants and needs. Part of this communication will involve all members being encouraged to express their feelings in a totally accepting way. Each needs to know, and consider, the others':

- needs;
- desires;
- attitude to power; and
- where they see the future evolving.

Similarly, their thoughts on the best course of action for each problem need to be fully canvassed and considered to finally arrive at a consensus.

Of course, not all families will need to work this way, as in any population there will be a full range of family situations and power distribution. Depending on the members' backgrounds and beliefs (more on this later), some families will be happy for the manager to be the sole decision maker in an autocratic operation. Increasingly, however, such situations are decreasing with the increased liberation of spouses and improving education opportunities. The family itself must work out, perhaps with help, the most suitable arrangement.

Some of the potential stress points include:

- a manager not considering the impact of decisions on each family member;
- a period of poor prices and weather/business outcomes, all giving rise to financial stress with few ways out of the situation; and
- the family members having very different objectives, goals and needs, and these may all have diverged since the business was set up.

The intergenerational transfer of resources is always a situation fraught with potential problems over fairness between siblings, though careful thought, discussion and professional help can lead to family harmony (see the succession section in Chapter 8). If parents remain in a business, it is hard to develop harmonious decision making with the son or daughter taking over. Different generations have different upbringings, agendas and beliefs that seldom match. This is an age-old problem existing in all parts of modern society where change is considerable. Integrating with daughters/sons-in-law may not be easy and requires a benign approach by the older generation moving to a different phase of their lives.

The list of potential family stress points is almost endless. But many families do work well and meet their objectives with satisfaction and harmony. Success must depend on an awareness of each person's:

- strengths;
- skills; and
- weaknesses.

This applies to all individuals in a family, so that they can pass on a problem to the person best able to handle it. Goal setting and time management are all important. Tasks not completed in time lead to enormous stress, in biological production in particular; but also in most other production systems, time waits for no one with, for example, the weeds seeding, the rain moving on, the market peak passing, no matter what the manager does. Thus, priority lists and realistic time estimations are important aspects to decision making in all business types.

Many believe that democratic decision making is important, if not efficient in the time taken. Involvement leads to commitment. And particularly in the most difficult times, it is important for all members of a family to have hobbies and recreational activities, including being part of the local community, not only for enjoyment, but also to help keep life in perspective. The old saying 'all work and no play makes Jack a dull boy' still has credence in more ways than one.

Despite the best of intentions, dysfunctional families do exist. The observer and the consultant must be able to recognize this and carefully help resolve the situation. Of course, in some cases, the situation is irretrievable, in which case the best course might be a major reorganization. A dysfunctional family can often lead to poor decision making and the non-achievement of goals and objectives for all members of the family. In these cases, it may be prudent to suggest professional help be called in, though this is difficult to achieve if the family does not recognize that they have a problem. Many organizations, and similarly professional counsellors, are available to help. Much literature is also available (search online, talk to community leaders, etc.).

The family problem may relate to the married couple, or involve the children as well, and even the previous generation. Each situation will require different expertise to help. A counsellor should be carefully selected to suit the personalities and beliefs of the people in conflict, so the discussions and contacts between families and professionals before selection are important. Most counselling involves a series of steps, which are followed through with the help of the professional. Seldom will a professional simply provide advice and instructions as solutions must come from the participants themselves. The professional will guide such resolutions and provide support. Usually an imposed solution simply does not work.

The steps often involve the following.

1. The individuals are asked to list out the good points in each other.
2. They are written down before telling the list to each other in a very positive way.
   Then, hopefully, the following is possible:
3. Each of them lists out the actions which she or he believes harms the relationship.

These steps require the participants to really listen to each other and absorb the feelings. This is where the counsellor can play an important role, because very often people do not want to hear about the 'not so good things' and brush them off. A real understanding of the other members' concerns and feelings is the first step to resolving issues. An acceptance of the 'no blame' approach is critical to the next step.

This involves taking responsibility for the concerns, and working out solutions that are mutually acceptable. Resolution is under way. Even without counselling, these open communication procedures can be practised. In either case, key words are:

1. 'Listen'.
2. 'Learn'.
3. 'Set aside time' to do both.
4. At all times practise remaining 'cool, calm and collected'.

This is where a third party can help make real progress.

Of course, counsellors will have their own variations on this basic process and resolution. And each family will require a slightly different approach, and the process will be dynamic taking its own individual course. Some families may prefer to tackle the situation by themselves, and certainly there are many good resources available. The local general medical practitioner may well have suggestions.

The emphasis in this discussion has been on tackling dysfunctional families. But families fall on a continuum of dysfunction with few being totally harmonious from a management point of view. Thus, the seemingly well-adjusted family may well obtain benefit from going through the procedure suggested.

Bruce the consultant has firm views that a consultant can be an important third person in helping families better integrate themselves into the totality of the business and the family. He has observed that dysfunctional families have a major impact on the business's success. And particular life-stage events can be concluded successfully with a consultant's help.

Two stages of notable importance are when a family member, most likely a son, returns to live and work in a business. Usually significant changes are necessary to provide the income to support an additional person, and often a shift of power must eventually be organized.

The other important stage is when the parents are ready to retire. But planning for this must start very much earlier and can benefit from the consultant's expertise, especially if she or he knows the family and the personality of its members.

Margrave's family was certainly not dysfunctional. Both Margrave and his wife (perhaps we can call her Margaret) were born in a middle-sized rural town on an idyllic coast. In Margrave's case, the family business was not far away and so it was convenient to take the trip into town. And later on Margrave's family, now including two boys, bought a house in town where they all lived so that the children would have easy access to education. Margrave then travelled daily to the business where most of the staff lived, ensuring proper cover. Besides, each staff member had his or her area of responsibility. Initially Margaret helped a lot on the business, and was involved in decision making, but once the move to town occurred the involvement was much less. Besides, Margaret, who is a university graduate, soon took up further distance education and had many challenges to occupy her mind.

The boys in their earlier life made full use of the business as an exciting playground, and in later years worked on the business carrying out jobs like cultivating large areas for growing winter feed. They were hungry for the good pocket money on offer and were happy to work long hours on this boring job.

It is clear the family influenced decisions regarding their living situation with education being paramount, but due to the later separation of their lives from the physical location of the business they were not heavily involved in the immediate decisions. And with Margaret developing her own professional expertise, her involvement was similarly somewhat divorced from day-to-day decisions.

And now the boys have largely left home having been through various training programmes leading to one being a project manager and the other involved in survey work. Neither has a strong desire to become involved in rural businesses and will probably make their way in urban life. But who can see the future?

In Hank's case, the family had a significant influence in their progression through life stages. Hanna was born in a rural area, as was Hank, so rural life was part of the family tradition. Further, both had strong agricultural and rural ambitions giving rise to clear objectives about creating a strong business, and in that they started with very little, they have been most successful. Clearly, shared views on their direction in life make the sacrifices and commitments so much easier.

As happens in families, the children were an important influence on resource allocation both from a daily living point of view, and from an educational view. Both children were encouraged to follow through with tertiary education and achieved these goals successfully (one graduated with a degree in Horticultural Commerce, and the other has completed diplomas in hospitality management).

Now, while their daughter is currently at home before departure to another country, it will soon just be Hanna and Hank at home. This completes the standard life cycle requiring Hanna to alter her objectives. This is not always easy because for so long the family objectives were very clear, with business and family being dominant in their struggle to develop resources and net worth. Hopefully, soon the re-evaluation will be complete setting forth new directions for the next stage in their lives.

The Influence of a Spouse

By far the majority of SME businesses will involve a partnered couple, and by far the majority of the spouses will be female. Thus, anyone involved in working with managers to improve their management, or even to just observe what is happening in a business, must consider the spouse's personality, intelligence, objectives, as well as family relationships.

Feldman and Walsh (1995, p. 39) noted: 'This division of power, control and skill shapes the decision process.' Clearly a spouse can be a powerful addition to the management team, both in a decision sense and in a business work sense, for an extra hand can be crucial in some tasks. Just how much a wife/partner becomes involved in business operations varies enormously across individual situations and cultures. In Queensland, for example, it was found (Rickson and Daniels, 1999) that:

- 51% of husbands believed they made decisions by themselves; and
- a further 30% believed that they tended to have more influence on decisions relative to their wife.

In contrast, for household equipment decisions:

- 49% of wives believed the decisions were shared equally;
- a further 33% believed they had a greater influence than their husband; and
- the remainder of the wives believed they made the decision on their own.

On the other hand, if you ask the husband, they believed that 25% of the wives made the decision on their own.

The figures for other decision types range around those quoted above and so it is clear that, overall, the managers do tend to have a greater influence than their wives in decision making, but this does vary from case to case as shown, for example, by 19% of the husbands believing decisions are shared equally.

Each family works out what they are comfortable with regarding the lead person in each decision. Of course, where the relative responsibility is in dispute, problems may arise. And these decision relationships are not static in that as a family evolves the decision-making responsibilities change. In times of financial stress it is often observed that a wife's interests suffer in favour of maintaining the business, and similarly the decision-making involvement suffers.

Another study (Sawer, 1974) found that wives who spent time finding information and data were more heavily involved in business decision making, as were spouses who spent time physically working on the business. In this situation, the wife probably made an important contribution to the decision involved. Sawer also found that the more children the couple had, the less involvement the female had in decisions, and there were similar negative relationships in decision involvement with respect to socio-economic status (income, business size, education, age, social participation).

A manager's adoption of innovations was also related to the spouse's decision involvement, so perhaps this involvement helped share the risk and the responsibility leading to the manager being prepared to take on the unknown at an early stage. It is also interesting to note that the wives in Sawer's study, on average, had 12 years of education whereas the husbands had 8 years, suggesting that many wives contributed through their better education.

In a study of 880 Kentucky businesswomen (Bokemeier and Garkovich, 1987), the researchers found that women who had been raised close to a business were more likely to be involved in business decision making and tasks and tended to have a strong self-identity regarding the business situation. This is a similar situation to managerial skill where it was shown that early business life was important in developing good business management skills. It is clear that all the work in the general population on the influence of early life on subsequent success appears to be similarly played out in SMEs.

It also appears that the household and the business are often in conflict over resources, and bargaining sometimes occurs over finding a compromise: 'If I can do 'this', then you can have 'that'.' This is particularly evident when the debt loads are high which in turn leads to stress. No one likes constantly increasing the debt load when conditions are difficult.

All the information presented makes it clear that the spouse is a very important member of the decision team in very many business households. In helping improve decision skill, just as much attention must be given to the spouse, and similarly when introducing technological improvement it should be ensured that the information is available to the spouse who should also be involved in any discussions, and, where possible, demonstrations.

A wife's self-identity is usually enhanced by involvement in decision making and business operations. There can be nothing worse than living in a relatively isolated situation and being regarded simply as a household keeper, though usually a wife is heavily involved in leisure decisions and activities thus leading to some satisfaction (Fig. 9.2 suggests craft can be a source of fun). Family decisions usually also provide satisfaction and identity.

Overall, for success and harmony, it is important to evolve good decision systems with a spouse contributing towards many aspects through discussion and resolution. Furthermore, a spouse can be actively involved in the decision steps from finding out information, sorting out goals, estimating costs and returns and resource requirements, and finally helping make the decision to proceed. It is interesting to note that on average women probably work longer hours than their male partners. One extensive US study (Kim and Zepeda, 2004) found that women work on average 13.8 hours per day over 7 days, whereas their partners worked 13.5 hours. This includes household tasks.

Working in a harmonious and beneficial way as a wife/husband team assumes that the marriage is successful. This is not always the case as shown by contemporary statistics (in the USA, something like 50% of first marriages end in divorce, 67% of second marriages and 74% of third marriages). The rate in different countries varies markedly, as does the number of official marriages per 1000 population. In New Zealand, the rate is as low as 14.6% and lowering with many de facto relationships. While it is not always easy to predict which couples will evolve a successful marriage, research does give some pointers. Understanding the personality traits which lead to success is important as it gives an idea of which couples are unlikely to be able to improve markedly, and in the other cases, which characteristics need targeting. Of course, helping dysfunctional couples is a specialist task, so any attempts in this area must be treated with caution.

One study (Bentler and Newcomb, 1978) found that personality was a successful predictor

Fig. 9.2. Spouse involvement can take many forms – in this case, another source of income as well as community involvement.

of a good marriage and, therefore, whether a couple might be able to sort out their difficulties. Similarly, Kelly and Conley (1987) found that the neuroticism (anxiety) of both the husband and spouse were important predictors as well as the impulse control of the husband. This study followed 278 married couples over 45 years. Twenty-two engaged couples did not make it to marriage of the 300 starting in the study.

Many studies have found similar results providing confirmation that the personality of the couples is important to a good relationship and, therefore, an increased likelihood that they will make good business decisions. For example, Caughlin *et al.* (2000) used a 13-year longitudinal study in which they measured aspects of the personality of married couples to assess correlations with marital satisfaction. The two most important traits they used were 'trait anxiety' (apprehensive, tense, emotionally labile, suspicious, shy and undisciplined) and 'negativity' (e.g. spouse showed anger or impatience by yelling, snapping or raising voice).

A negative relationship between trait anxiety and both marital satisfaction and communication was found.

A positive relationship existed:

- between an individual's trait anxiety and negativity;
- between the wife's trait anxiety and the husband's negativity;
- between both the husband's and wife's marital satisfaction and the spouse's negativity; and
- the husband's negativity was related to the wife's marital satisfaction.

It was also clear that spouses were directly affected by their partner's emotional make-up. The implication from all these studies is that working on improving aspects of couples' personalities is likely to improve decision efficiency as well as marital satisfaction. As noted in the following chapter (Chapter 10) on skill improvement, there is good evidence to suggest that personality (management style) can be altered given good support, a strong desire to change and skilled counselling.

But for Margaret and Margrave, it appears that marital disharmony was certainly not an issue in their lives. Margrave could devote all his energies to making good decisions knowing home support was always available. Margaret has been, and is, involved in a range of jobs, and besides a university bachelor's degree, also has training in school teaching, business administration and also completed a university master's degree. Some of this training was through distance education reflecting a determination to succeed. Her employment has involved professional jobs in both rural and urban occupations, and still does. All this experience and training has enabled Margaret to provide excellent support in all decision and analytical problems related to the business. This is not uncommon in Western rural life with highly intelligent spouses looking for outlets for their talents and for satisfaction in making important contributions to the community.

The team that many rural couples develop is powerful once the children are less demanding. And prior to this stage, the assistance and example provided to the children are similarly powerful and motivating. And of course there is the extra pair of hands provided at crucial stages. The life cycle described here is an example of what the literature reviewed reveals.

And Hanna has contributed to the business development in Hanna and Hank's case. Hanna graduated with an Agricultural Science degree and was, and is, highly qualified to be part of the decision team. Now that their children are more or less independent, Hanna has become heavily involved in the recording and accounting side of the business and constantly produces critical financial reports. Prior to some health difficulties, Hanna was also involved in raising calves, a task which she loved and gained considerable satisfaction from knowing that the results led into valuable herd assets. Hanna has also contributed heavily to the rural community and over the last few years has been the president of the local golf club ensuring its continued place in the sporting and social life

Fig. 9.3. Contribution to the local community is often an important part of rural life and can take many forms.

of the district. Others might involve themselves in church activities (Fig. 9.3). Rural people are very aware that they depend on each other for so many parts of their lives. That is what makes a country.

Bruce the consultant is also under no illusions about the importance of a business spouse in the overall scheme of things. Bruce comments: 'A consultant must work with the spouse in most cases for 'she' has a major influence in what happens.' Sometimes the spouse is from a long-standing business family, who may well have provided resources to help the 'young couple', and consequently rightly believes she is an important part of the system who has considerable knowledge of the business. The consultant, logically, must deal with both members of the team.

The Origins of Objectives

As often stressed, a lynch component of managers and their families are their objectives which drive and underpin all decisions. It is crucial to understand both their nature and origins when working with managers for otherwise any discussion on possible changes is likely to be misdirected. Given the stimuli being received by the manager, such as next season's price predictions, the objectives should dictate the management response. To ensure an appropriate response the manager must clearly understand the objectives both s/he and the family hold. As Hobbs *et al.* (1964, p. 1) noted 'objectives set the goals, and strategy sets the path to the goal'.

As noted, there are many sets of questionnaires in the literature for assessing objectives including the one listed in Chapter 3 (which the reader may wish to review). In this literature, different terminology is often used. Examples include the terms 'values' and 'goals' as well as 'objectives'. However, when you think about it all three words can logically be used.

Every manager, and the family members, will have a set of basic values, such as a belief in being independent and secure, which they wish to adhere to. These give rise to their objectives. For example, wanting to be independent and secure probably leads to an objective of reducing debt. In turn, the goal may be to reduce the mortgage by 15% over the next 3 years. Thus, values lead to objectives which, in an operational sense, give rise to specific goals and targets which then lead to actual plans and forecast budgets.

You probably have a wide range of values which guide how you react to different situations. Similarly, managers and family members will have developed value sets resulting from their background and experiences. The set will be heavily influenced by the early family environment, and similarly the community in which they live, as well as the general culture encompassing both the family and community.

Examples of basic values include:

- the degree of security sought;
- the drive to conform;
- the level of power sought;
- the benevolence level;
- friendship;
- loyalty;
- forgiveness;
- helping others;
- politeness;
- attitude to equality;
- the respect required (community status);
- honesty and openness;
- anti-pollution by both natural substances and man-made substances;
- sustainable production and organic production values;
- leaving the business in a better state;
- independence;
- importance of innovation (creativity);
- valuing education for its own sake;
- balancing work, leisure and comfortable living conditions;
- working for an adequate living and conditions compared with maximizing;
- not working, in the case of Christians, on Sundays with respect to religious beliefs; and
- ensuring full use of abilities and resources as a national responsibility.

And so the list can continue.

Not all values will necessarily impact on business objectives, but in that a family business intrudes into most parts of a manager's life and times, most will impinge on operations at some time or level. Values such as honesty are all pervasive, and even values such as emphasizing education will have some influence on the objectives as it may, for example, be necessary to curtail business comforts to provide education funds.

In an important early review, Gasson (1973) summarized the wide range of possible values into four main categories. She called them:

1. Instrumental (e.g. business as a means to an end, such as income, security, congenial working conditions).
2. Social (e.g. being part of the SME business community, good relationships with employees, family relationships).
3. Expressive (e.g. self-respect, creativity, overcoming challenges).
4. Intrinsic (e.g. value of independence, enjoying outside work, family involvement).

Gasson reported on a range of business surveys and concluded that intrinsic values were particularly important to most managers, but that the manager's situation also correlated with the importance of the other main categories. For example, low-status managers particularly valued independence, and the social values were important to middle-status managers. In contrast, the higher-status large businesses had managers who valued instrumental values. They had moved into being serious businessmen/women striving for maximum profit.

However, it must be stressed that categories can be dangerous, and each manager has a unique background and situation that gives rise to her or his particular set of values, objectives and goals. And it is also worth noting that the research reported concentrates on the manager's values in contrast to the total family values, though in some cases the manager's values will reflect the totality of the manager and the family. In modern times any research should consider both the manager and his spouse, and possibly other family members too. As the years have passed the business spouse has become an increasingly important part of the management team.

Values, objectives and goals are seldom static, so anyone considering business, managers and business families must constantly review conclusions and associated family support. For many years business management studies assumed the objective was to maximize profit, and managers not reaching this nirvana would be regarded as irrational (some, for example, might find nirvana in fishing, see Fig. 9.4). In reality, it was the commentator who was irrational for few managers have a simple objective of maximizing profit.

Indeed, many might say today that most managers are rational; it is only the inability to define their goals that makes them appear irrational. This is probably not true in many cases because efficiency studies that relate one manager to the others show a significant proportion could improve output with the same resources or maintain output with fewer resources. Again, some of these conclusions probably stem from an incorrect assumption about what criteria to use, but certainly not all managers believe they are competent and rational as evidenced by the many managers who have employed consultants over long periods, which suggests the managers believe they are getting value for money.

Part of the dynamism of objectives is the changing situation managers find themselves in. Most businesses follow a kind of life cycle in that in their youth managers and their spouses are young and enthusiastic with values formed by their families and community. As the years pass the presence of children in the family often impacts on values, and then eventually, when the children leave and the years take their toll, values and objectives undertake further change.

Thus, what is right for one family will be wrong for another depending on their stage in life. With time managers tend to become more conservative and change less. And where a business has been in the family for several generations, the objective set is probably altered for the attachment to the land can be powerful, leading to sustainability and environmental issues that will often be stronger relative to a manager on a new business.

Values are partly determined by a manager's personality so given that it can change in earlier life (and also later where special efforts are made), values can similarly change.

Fig. 9.4. A manager with this resource bordering his business investment is truly lucky, and who would blame him if it influenced his objectives.

Other factors giving rise to change are extended family influences such as, say, a financial input by relatives to enable expansion, or a change in ownership structure (perhaps part of the transfer of the assets from one generation to the next). Similarly changes in the general business environment may lead to changes. If there is a major shift in, say, prices leading to the impossibility of paying back a family mortgage, it will mean objectives have to change. When an objective cannot be achieved the human condition usually leads to change creating objectives that are more realistic. If modifications did not occur, emotional turmoil in seeking the impossible could easily ensue.

Finally, it is important to accept that managers are seldom totally stable in expressing their values and objectives. Over, say, 2 years a manager might well make a rational decision over marketing expenditure in the spring this year, but be quite irrational next spring. The major reason, all other things being equal such as prices and costs, values and objectives, is the impact of emotions. Being human means we often react according to our emotional state. Perhaps the local managers' organization has voted the manager off the committee for no apparent reason other than the words of a rival, making the manager angry at his neighbours. In an irrational response he orders and applies too much advertising going well past the point of marginal cost equalling marginal return in an effort to show up his neighbours as bad managers. Managers are not machines and, like all humans, react to their emotions. We cannot always help it.

Atkinson *et al.* (2000) talk about the components of an emotion as being the following:

1. A subjective experience.
2. Internal bodily responses (e.g. a fright response).
3. Cognition about the emotions and associated situation.

4. Facial expression (you have all noted embarrassment).
5. Reactions to the emotion.
6. Action tendencies.

Of course, the good manager takes time to review an action relative to the objectives, and hopefully controls irrational action. Some people need help in maintaining this rationality.

Quantifying the Origins of Objectives

Introduction

To understand the origins of a set of objectives, and therefore consider how to possibly modify a manager's, and a business family's objectives, studying the factors influencing the objective set is important. This requires data. For students interested in researching objectives they should consider Garforth and Rehman's (2005) review of values, goals and objectives. In this they consider the alternative data collection methods. They all involve questioning managers in various ways.

Presented below is an analysis of data collected through a postal survey of over 2000 managers yielding 735 responses. While the data was collected to quantify the important factors in determining managerial skill (Nuthall, 2009), much of the data was also useful for exploring the factors leading to specific objectives.

The important factors

Listed below are the factors which logic would suggest are likely to determine objectives. How such factors influence the combined family objective set is another matter, which may or may not be a simple amalgam of each individual's objectives. This is somewhat of an unknown in a research sense. The relative dominance and importance of each individual will impact on the group objectives, which may well be somewhat fluid. Too much dominance can, and does, lead to resentment which eventually surfaces in a range of ways.

For an individual, while each case will be unique, the likely factors influencing objectives are as follows.

1. The parents', and possibly the grandparents', objectives and outlook on life.
   Along with most factors, the childhood environment will have a major influence on objectives. And similarly the genes inherited, if, in fact, a person's genetic make-up is important to the values and objectives. As part of the family influence are the family traditions (e.g. perhaps employing rather more of the local people than might be prudent as they have always employed from specific local families; being heavily involved in the local religious organization and its values). Also an important part of the parental influence will be the encouragement and general support both for school and other activities.
2. The manager's management style (personality) and intelligence, which are also, of course, caused, in part, by the genetic heritage.
   A person's ability to think through issues resulting from his or her intelligence will influence the objectives. For example, reasoning relating to management systems and their impact on the environment will affect both the view on acceptable systems, and the goals resulting from the reasoning. The same comments can be applied to personality, so, for example, someone with high anxiety levels is likely to have a stable income as an objective in contrast to a maximizing income.
3. The manager's age, sex and education will influence the objectives.
   Age, clearly, is part of the life cycle and also age itself can alter how a person sees his or her place in the world; and certainly the manager's gender impacts in that the hormonal system influences how people view life and what is important. Education to higher levels can also influence how people think about their existence and what might be crucial outcomes in their life. The type of school(s) attended, and the specific mix of teachers experienced, can have a powerful influence, including

whether the schools were rural- or urban-based.

4. The resource situation a manager finds himself in will also impact on the objectives if for no other reason than they determine to a certain extent what is physically possible. Often people adjust their goals to something which is practical and feasible. But on the other hand, some people are very determined to achieve certain goals and set their sights very high, and a subsection of such people will make it happen (and, no doubt, some have very good luck helping the cause, such as a long run of favourable seasons and prices). The resources available include the quality and size of the business and its attributes, and the lack of a high debt load. The number and quality of the employees also impact what is possible.
5. In a similar vein, the manager's inherent managerial skill can impact on the objective set for it will influence both financial and other success and, therefore, what is practically feasible.
Associated with managerial ability will be the experiences a manager has had, the early life and encouragement offered by his or her parents, and so on, as noted in Chapter 3.
6. A manager's inherent enjoyment of working with the business assets, such as the machines, buildings, soils, plants and animals, can be a powerful factor in objectives and in many cases is the reason he or she will accept a lower income relative to what might be possible from selling up and working in a city.
Similarly, their interest in the natural environment (evidenced by the many SME rural managers who have promoted, and fenced off, areas to allow indigenous species to flourish) and sustainable production that will improve rather than deplete the environment are all factors in setting objectives and goals.
7. And finally, is a manager's interest in people per se.
This factor will be partly determined by the manager's personality, and the family background, but it needs mentioning separately for some managers will put special effort into helping others and their community as a specific objective. An example would be the manager who devotes many hours to manager politics and lobbying politicians to enhance the managers' general cause. This usually comes at the cost of income and an enjoyment of business for its own sake.

Some quantification

The data from the manager survey was used to develop linear equations to reflect the importance of each variable for which data was available. This was carried out for each factor, or component, of the managers' objective set. In the surveys reported in Chapter 3, five distinct factors were numerically isolated. In the present survey, these same factors emerged from the factor analysis, but also a sixth factor was apparent. Somewhat by chance, the sample had in it managers who were somewhat reluctant managers. These were people who did not enjoy business but had no other qualifications or opportunities enabling them to leave the business (or perhaps some were rather timid). This list of objective components is as follows.

- Work enjoyment – to simply enjoy the process and environment.
- Wealth – to obtain wealth through both income and asset value increase.
- People – to enjoy contact with people including business workers, other managers, and family including being able to set up children with assets.
- Leisure – organizes business to ensure sufficient leisure and holidays, community activities and time to try new things.
- Risk – uses business systems that ensure risk levels are kept low.
- Stoicism – continues business as an objective simply as he or she has no alternatives.

The totality of any manager's objective set is made up of proportions of these components, each of which has a different degree of importance for any particular individual.

The variables that entered the equations were the following (unless otherwise stated, most variables took a 1–5 scale reflecting the degree of truth in the statement):

Age the age, in years, of the manager
Ed highest level of education (1 = 'primary', ..., 5 = 'three or more years of tertiary')
Lab the number of employees on a business
Gen the gender of the manager (1 = 'male', 2 = 'female')
Siz a measure of the economic size of the business, being the product of the area (ha) and a code reflecting the intensity of the business type (6 = 'intensive horticulture', ..., 1 = 'extensive sheep')
St1 management style factor: concern for correctness (anxiety)
St2 management style factor: thoughtful creator (openness)
St3 management style factor: conscientious planner (conscientiousness)
St4 management style factor: community spirit (extroversion – community)
St5 management style factor: consultative logician (extroversion – family and friends)
St6 management style factor: benign manager (agreeableness)
Ee1 childhood early experiences: helped with business jobs
Ee2 childhood early experiences: reason for decisions explained
Ee3 childhood early experiences: listened to discussions on finances
Ee4 childhood early experiences: listened to discussions on technical matters
Ee5 childhood early experiences: asked my opinion on decisions
Ee6 childhood early experiences: wanted to know reasons for decisions
Es1 as a child, encouraged to use imagination to find solutions and methods
Es2 as a child, encouraged to improve observational skills
Es3 as a child, encouraged to 'get along' with friends and relatives
Es4 as a child, encouraged to learn agricultural knowledge
Gen number of generations on the particular business
PO parents' objectives are different
GO grandparents' objectives are different
SM self-scored managerial ability (1 = 'poor', ..., 10 = 'excellent')
EM managerial ability calculated from the productivity and other data (%).

The equations were calculated using all these variables together with a number of other similar ones. They were all the data available on the variables logically expected to influence a manager's objectives. Given the results, variables that had a probability of greater than approximately 20% of being no different from zero were dropped and the equations recalculated. These are presented below with the significance figures of the remaining variables given in parentheses below the coefficient, and the standardized coefficient in the square brackets.

These standardized figures allow a comparison between variables to assess the importance of each as they neutralize the differences in the units of measure. The equations were all highly significant, and the percentage of the variance explained by the equations ranged from 10% (stoicism) to 31% (risk). These figures are not particularly high, indicating other factors are also involved including the variables listed as potentially important but for which no data were held.

Clearly further research needs to explore details of the parents' influence, and similarly the impact of culture. Questions relating to these variables were not available in the questionnaire. Details of the equations are as follows.

For work enjoyment:

$$\text{Work enjoyment} = \underset{(0.02)}{0.95} - \underset{\substack{(0.13)\\ [0.06]}}{0.06\text{Age}} - \underset{\substack{(0.01)\\ [0.10]}}{0.76\text{Gen}} + \underset{\substack{(0.07)\\ [0.08]}}{0.08\text{St2}} + \underset{\substack{(0.00)\\ [0.17]}}{0.18\text{St3}}$$

$$+ \underset{\substack{(0.04)\\ [0.08]}}{0.1\text{Ee1}} + \underset{\substack{(0.00)\\ [0.14]}}{0.14\text{Es3}}$$

The most important variables are age and the constant (which could well be explained by some of the unavailable variables), and this is followed by conscientiousness (St3), early encouragement to get along with people (Ee3), and then gender. This suggests personality and parental influences give rise to work enjoyment.

For wealth:

$$\begin{aligned}\text{Wealth} = \underset{(0.45)}{0.29} &- \underset{\substack{(0.08)\\ [0.07]}}{0.05\text{Lab}} + \underset{\substack{(0.04)\\ [0.09]}}{0.09\text{St1}} + \underset{\substack{(0.00)\\ [0.17]}}{0.16\text{St2}} + \underset{\substack{(0.00)\\ [0.13]}}{0.13\text{St3}} - \underset{\substack{(0.03)\\ [0.09]}}{0.90\text{St6}}\\ &- \underset{\substack{(0.14)\\ [0.06]}}{0.06\text{Es3}} + \underset{\substack{(0.15)\\ [0.06]}}{0.05\text{Es4}} + \underset{\substack{(0.00)\\ [0.14]}}{0.12\text{Ee5}} - \underset{\substack{(0.15)\\ [0.07]}}{0.06\text{Ee6}}\\ &+ \underset{\substack{(0.11)\\ [0.07]}}{0.05\text{PO}} - \underset{\substack{(0.02)\\ [0.11]}}{0.18\text{SM}}\end{aligned}$$

The source of the wealth objective is rather more complicated than work enjoyment. The constant is not nearly as important, and furthermore, with a significance of 0.45 the chance of it being no different from zero is high. The three most important variables are openness (thoughtful creator) and conscientiousness from the personality variables, and the parental influence of encouragement to be involved in decision making at an early age (Ee5). The other variables also help determine the importance of the wealth objective but are less important. The desire for wealth clearly has many components.

For people:

$$\begin{aligned}\text{People} = \underset{(0.78)}{0.10} &+ \underset{\substack{(0.05)\\ [0.07]}}{0.05\text{Lab}} + \underset{\substack{(0.16)\\ [0.06]}}{0.05\text{Age}} - \underset{\substack{(0.02)\\ [0.10]}}{0.09\text{Ed}} + \underset{\substack{(0.00)\\ [0.30]}}{0.30\text{St2}} + \underset{\substack{(0.00)\\ [0.22]}}{0.23\text{St4}}\\ &+ \underset{\substack{(0.00)\\ [0.18]}}{0.18\text{St5}} + \underset{\substack{(0.12)\\ [0.06]}}{0.06\text{Es3}} - \underset{\substack{(0.17)\\ [0.06]}}{0.04\text{Es4}} - \underset{\substack{(0.15)\\ [0.06]}}{0.05\text{Ee2}}\\ &+ \underset{\substack{(0.09)\\ [0.09]}}{0.06\text{Ee6}} + \underset{\substack{(0.03)\\ [0.08]}}{0.06\text{GO}}\end{aligned}$$

The desire to interact with people as much as possible similarly has many constituents. The personality variables (St) again feature, with openness and extroversion being particularly featured (St2, St4 and St5) as would be expected. Education also impacts through, perhaps, giving confidence.

For leisure:

$$\begin{aligned}\text{Leisure} = \underset{(0.01)}{0.49} &- \underset{\substack{(0.00)\\ [0.12]}}{0.12\text{Age}} + \underset{\substack{(0.00)\\ [0.12]}}{0.11\text{St1}} + \underset{\substack{(0.00)\\ [0.13]}}{0.13\text{St3}} + \underset{\substack{(0.00)\\ [0.25]}}{0.25\text{St4}}\\ &+ \underset{\substack{(0.00)\\ [0.12]}}{0.15\text{Ee1}} - \underset{\substack{(0.03)\\ [0.09]}}{0.06\text{GO}}\end{aligned}$$

Once again the personality variables are strongly linked to the desire for leisure (particularly extroversion, which perhaps expresses a need to take time off to interact with people in spare time and holidays), but also age with older people being less interested in leisure.

For risk:

$$\begin{array}{llllll}
\text{Risk}=-0.98 & -\ 0.20\text{Age} & +\ 0.11\text{Lab} & +\ 0.30\text{St3} & +\ 0.07\text{St4} & +\ 0.03\text{EM} \\
\quad\ (0.00) & \quad (0.00) & \quad (0.00) & \quad (0.00) & \quad (0.06) & \quad (0.00) \\
 & \quad [0.21] & \quad [0.17] & \quad [0.17] & \quad [0.07] & \quad [0.34]
\end{array}$$

The large negative constant suggests people are inherently risk preferers, but with this being strongly negated by conscientiousness (St3) and age. Thus, as managers get older they tend to get more conservative as is often observed in the field.

For stoicism:

$$\begin{array}{lllll}
\text{Stoicism}=-0.82 & +\ 0.09\text{St1} & -\ 0.16\text{St2} & -\ 0.06\text{St3} & +\ 0.01\text{EM} \\
\qquad\quad (0.00) & \quad (0.03) & \quad (0.00) & \quad (0.10) & \quad (0.00) \\
 & \quad [0.09] & \quad [0.16] & \quad [0.06] & \quad [0.17]
\end{array}$$

This rather less important objective, which is more about an accident of history than an objective, is largely explained by a manager's personality with openness being particularly important as is managerial ability (EM). Managers with a reasonable ability stick with the business despite not enjoying the occupation, and managers with a high level of openness, on the other hand, tend not to be stoic and probably tend towards giving up the business.

Overall, while quite a few variables clearly give rise to the objectives held by managers, it is clear that their management style (personality) is a major factor. This set of variables is conferred on a manager by the manager's parents and early environment.

Age, education and business size also have some influence as you would expect. Finally, some of the early experience variables – 'helped with business jobs' (which implies the manager lived on a business from an early age), 'decisions explained', 'party to discussions on family finances' and 'enquired about the reasoning behind decisions' (again implying early-age business experience) – were also significant.

Thus, to change a manager's objective set involves working with the manager over changing his or her personality, and similarly the impact of some of the early experiences. This raises questions of what is ethical. Many would conclude that an advisor or consultant is empowered to work with a manager to improve the attainment of the stated objectives, not to change the importance of each objective.

Indeed, in the early days of consultants working with managers many were later criticized by the managers for pushing them into intensive business operations with a view to improving profit. This often worked for a while but when conditions altered, considerable stress resulted from both disastrous downturns and sleepless nights from the high risk, let alone the family reaction to less leisure time. More latterly, advisors and consultants accept manager's objectives for what they are rather than assuming everyone wishes to simply maximize profit or has the same objectives as the consultant.

Many would consider an ethically acceptable approach is for a consultant to work with a manager, and his family, in reviewing their objectives to see if they are indeed what they want. It is also ethical to help a manager whose emotions often override sensible decisions to maintain a more rational approach when making decisions aimed at achieving their objectives. This support requires training the manager to consider decisions calmly and rationally in contrast to 'spur of the moment' reactions. This will likely require constant contact between the people involved as decisions are made in the initial phases or training.

The action steps to be followed in reviewing a manager's objectives are as follows.

1. Set out what the manager and family believe are the objectives. This will involve:
   a. Use of the questionnaire on objectives.
   b. Observation of the actions taken by the manager.
   c. Talking to the manager and family and asking them to come to an agreement on the objective list through reviewing both the questionnaire results and the observations.
2. Discuss with the manager and family whether they believe the objective list and their priorities are actually what they want. The process followed in (1) will frequently start the participants thinking whether in fact their list is appropriate, especially as it is unlikely they have ever before enunciated what they want from their business.
3. Review whether the revised list is actually what the manager and family want after allowing time for the participants to mull over the conclusions. Often this should be a constant process as conditions and life-cycle stages change. Objectives are dynamic.
4. Make plans and organize support systems appropriate to the revised objectives.

Locus of control

As noted in the Introduction to this chapter, a manager's LOC may be hindering achieving the objectives and so working with the manager to help develop a more appropriate locus can be beneficial. It was noted in Chapter 3 that a manager's locus was heavily dependent on his management style. But, similar to the objectives, the origins of his beliefs may involve further variables. To test this idea, the same set of data used for the study of objectives was used to develop an LOC equation. The equation with only the significant variables included was:

$$\text{Locus of control (\%} = \underset{(0.00)}{52.98} + \underset{\substack{(0.00)\\[0.11]}}{0.85\text{Ed}} + \underset{\substack{(0.00)\\[0.19]}}{1.53\text{St1}} - \underset{\substack{(0.00)\\[0.32]}}{2.63\text{St2}}$$

$$+ \underset{\substack{(0.01)\\[0.09]}}{0.70\text{St3}} - \underset{\substack{(0.09)\\[0.06]}}{0.51\text{St4}} + \underset{\substack{(0.19)\\[0.05]}}{0.41\text{St5}} + \underset{\substack{(0.08)\\[0.06]}}{0.41\text{Es4}}$$

$$+ \underset{\substack{(0.00)\\[0.13]}}{0.81\text{Mis}} + \underset{\substack{(0.00)\\[0.19]}}{0.12\text{EM}}$$

where the variable names are the same as used in the objective equations with the addition of Mis, which is a 1–5 score on how quickly the manager takes up a lesson from experiences. The equation was highly significant, as were most of the variables. The equation explained 32% of the variance showing that other factors are also involved.

It is again clear that a manager's management style (the St*x* variables) is very important in creating his locus, particularly the openness character (St2), but also education, an ability to quickly learn from experience (Mis) and managerial ability. The latter is somewhat obvious and unhelpful, whereas working on improving the uptake of the lessons from experience, and similarly the openness character in management style, may well lead to a more appropriate locus.

Entrepreneurship

Some managers aspire to be an entrepreneur; others are content to follow the well-proven paths of conventional management systems using what productive resources they currently command. However, it is useful to comment on the factors associated with entrepreneurship for some managers may benefit from reviewing their objectives relative to being more adventurous. Clearly, objectives and entrepreneurship are highly related in that if a manager has maximum profit as a high priority then acquiring the characteristics of entrepreneurs may well be beneficial. But can they be achieved?

A search of the literature shows that while entrepreneurship is widely recognized as an important phenomenon, there is a paucity of work explaining and proving it can be fostered. However, the research does show that entrepreneurship is related to a number of personal characteristics where it is assumed that entrepreneurship is reflected by business success.

It is commonly suggested that the following factors are positively correlated to entrepreneurship:

- the need for achievement;
- power; and
- leadership.

Also important is the

- degree of personal ownership of the business with the higher the proportion, the greater the entrepreneurship.

In rural business the manager's ownership level is frequently very high. The tests used in assessing entrepreneurs have, in addition, shown a positive correlation with:

- the need for influence;
- verbal aggression; and
- hostility.

Stewart *et al.* (1998) showed that entrepreneurs were higher in achievement motivation, risk-taking propensity and preferences for innovation. They talked about the profile of an entrepreneur as being a driven, creative risk taker who plans heavily. Clearly they have goals of profit and growth. What is noteworthy is that risk taking is heavily linked to the anxiety personality trait and this trait can be influenced through proper training and support. To this end, entrepreneurship can be fostered.

In an extensive review, Judge and Ilies (2002) also conclude that performance motivation is highly correlated with anxiety and conscientiousness, and that the other personality traits (openness, extraversion and agreeableness) are also correlated, but not so strongly. Taken together the five traits were 49% correlated with the motivational criteria they used.

In a specifically rurally directed piece of research (de Lauwere, 2005), it was concluded that all of the following affected entrepreneurship positively, but that 'love of ease' and passivity had negative impacts:

- self-criticism;
- leadership;
- creativity;
- perseverance; and
- initiative.

Overall, like motivation, entrepreneurship is clearly highly related to management style (personality) with the manager's goals also having an influence. A manager interested in, say, sustainability, in contrast to maximum shorter-term profit, can have all the right personality traits, but would not be labelled an entrepreneur in the conventional sense. For a manager reviewing their goals and methods, they might well consider counselling and support to modify their personality traits if they conclude they wish to be more entrepreneurial.

For the case study managers, both regard themselves as entrepreneurs, though Hank has questioned his approach more latterly in this respect. On further reflection Hank did concede 'compared with most managers I guess I'm quite entrepreneurial'. He is certainly that for he has grown the business in leaps and bounds over the years and now has two separate dairy businesses which are both growing in output. When asked the reasons for his entrepreneurial approach, Hank was somewhat lost for words. He did note, however, that a favourite uncle probably influenced him with his stories of expansion and business growth. He aspired to follow in these footsteps.

Margrave comments that he became an entrepreneur out of necessity. This might be said about all entrepreneurs in that they either have to be venturesome to survive, or their management style and objectives drive them to follow the entrepreneurship path. When first taking over the family business, Margrave experienced a major downturn in the agricultural 'business climate'. Subsidies, incentives and agricultural protectionism were all suddenly removed by the government at the same time as prices were weak. The upshot was he had to branch out into all sorts of additional activities to be economically viable. Examples include rabbit meat production and wood supply activities. In the end, none of these initiatives was sufficient and the business was sold to eventually be replaced by other arrangements. Despite this, Margrave believes he is an inherent entrepreneur through his keenness to try new things: '[It] makes life interesting, gives life an edge ... I always enjoyed walking out of the house to start work each day and have this very high motivation. Part of my personality I guess.' Yet his father was not an entrepreneur, perhaps because he went through the Great Depression, and this created a cautious approach to debt and unproven ventures: 'My father even straightened out old nails to save expense.'

Concluding Comments

Anyone studying the human factor in business management must recognize that the manager and family must base their decisions on a set of objectives which accurately reflect their wants and needs. Objectives are very much part of the human side of business, and these involve considerably more than the simple material outputs including cash profit. The classical assumption of maximizing profit is a total misnomer except, perhaps, where the owners of the business are totally divorced from the physical location and operational management of the business.

Assessing the objectives is a human problem, as is checking that the professed set is correct. In that SME businesses worldwide are largely family-owned and family-managed, assessing objectives must involve the whole family. Each family will be unique in its make-up and objectives, and in its cohesion. A successful family will consider each member's wants and needs and communicate successfully, leading to democratic conclusions accepted by all.

Clearly there is competition for resources given the business and family are intimately related so resources devoted to the business means the family will need to go without, and vice versa. However, some outputs from the business will directly provide family satisfaction thus representing a complementary situation. An example is the space, views and peace provided.

One of the first things an observer must note are the 'business's' objectives for it is not possible to judge success without this yardstick, and therefore what might be improved. Part of judging the situation is noting the family cohesion with particular emphasis on the relationship between spouses. Where disharmony exists, everyone can gain if this can be approached and lessened, perhaps with outside professional help.

Business involves many stresses and in the bad patches disharmony can have a major impact on efficient management. In a 'business difficult' summer, for example, where the manager is working very long hours without family support, a break point can be reached. Furthermore, each family will go through a business life cycle with some of the stages being potentially more stressful than others. When the children need secondary or tertiary education the stress on available funds will be extreme, especially if this occurs in a low point in the inevitable price cycle. Education may well involve the children living away from home, thus further creating stress and a family dynamic that is totally changed with just the manager and spouse being left at home with, perhaps, time on their hands.

In most cases a spouse will be female, but whatever the case, 'marital' harmony is an important aspect of efficient resource use, particularly where the spouse is heavily involved in both working on the business and in decision making. An observer must talk to both partners, assess whether improvements are possible and call in what help is regarded as necessary provided the request comes from the spouses. The observer needs to be able to move the participants to the point where they realize there is a problem that might be retrievable. Or perhaps enlist the help of others in the family or community. Assessing whether improvement is possible partly involves assessing the personality of the participants for the available research clearly shows the link between personality and marital success.

Also important are the basic values and backgrounds of the spouses. It was clearly noted that core values lead to objectives which in turn create goals to be achieved. If spouses have quite different values it is hard to achieve cohesion unless they have benign personalities that enable major compromises.

Another important aspect of personality is how a person reacts to emotions. Being human, managers have emotions that sometimes override rationality and so it is important for a manager to recognize this and take time to consider particularly important decisions. A stable and helpful family situation helps attain emotional stability and appropriate decisions. Human beings can get depressed, and some extremely so. But family support, and professional help in some cases, can negate this depression and aid rational decision making. The state of any depression is not simply black or white, but rather a condition that everyone suffers at some stage. There is a continuum from euphoria to the deepest depression but with most having simple 'down periods'. In that managers are dealing with many uncontrollable factors, if they all turn down at the same time anyone would get somewhat 'down' in their feelings. Major problems can occur, however, when both the weather and markets are unfavourable for a long period.

Finally, if an observer understands why a particular set of values, objectives and goals is held it is much easier to work with the manager and family in checking their appropriateness and possible modification. The research available suggests personality has a major influence on objectives, and this in turn is dependent on parents, the early environment and community. Assuming a family wishes to check and possibly change their objectives, and perhaps modify their behaviour to provide better family harmony and, thus, more efficient decision making, consideration must be given to the ethics of influencing them using a range of methods.

The key is to let the family drive the operation following totally open discussions and conclusions. Better attainment of the truthfully held objectives is the goal, and where attained, provides considerable satisfaction to all involved.

10 Methods of Improving Managerial Ability

Abstract

This final chapter covers all the processes that may be useful when improving a manager's decision skills. Overall, it is useful to have benchmarks to assess a manager and the associated business, providing status judgements. Critical to success are personality, a reasonable intelligence, training, experience, technical knowledge, self-critical skills and a willingness to change. There is always much to be learnt with a review of learning processes presented, but 'one system does not fit all'. Learning is constant with feedback an important component. Becoming a better manager is an investment akin to any other investment. Research shows the time and cost involved usually provide a handsome return. The list of topics that might be studied, either formally or self-taught, is reviewed (e.g. record keeping, memory development, creativity, forecasting ...). Personality enhancement is also possible. Methods of up-skilling are discussed (e.g. experiential, mentoring ...) as are the approaches that are helpful. The case managers and consultants make their views clear, and, finally, a discussion covers the all-important 'gender issues' topic as an example of other human characteristics that may cause potential issues.

Introduction

The main reason for studying managerial ability is to consider ways of improving the SME manager's managerial skill, though an understanding of managerial improvement can also be useful when considering the impact of agricultural policy initiatives. The purpose of this chapter is, therefore, to consider the techniques that will improve a manager's skill no matter at what level they start.

As every manager currently exhibits a particular level of ability, a set of methods that can initiate improvement in all situations is required and is highly desirable. Some managers will improve more than others both due to their starting point and inherent ability. Each starts with a certain potential as defined by their genotype, and their early environment and experiences (phenotype is the genotype plus experiences in total). While the genotype is fixed, additional training of various kinds can change and improve the impact of their experiences. Fortunate managers will have an appropriate genotype (intelligence, personality, etc.), and appropriate early experiences in the form of family life, education, challenging situations, encouragement and training courses. These all lead to skill, curiosity, confidence and self-esteem. Managers without these advantages must work at compensating their situation with the support of all the facilities that are available.

Deciding on a manager's ability depends on the use of benchmarks including technical and economic standards. The latter are easy to obtain and calculate. What is not so easy is the testing of a manager's aptitude and managerial style with particular reference to their personality. While general tests exist, tests for attributes like self-esteem, motivation and people skills specifically for managers are not yet generally available. Thus, an assessment is dependent on observations of a manager at work. Such assessments rely on the concurrent observation of other managers to help form benchmarks of what is possible. For technical and financial figures, various techniques are available including data envelope (Dhungana *et al.*, 2004) and stochastic frontier analysis (ten Raa, 2022) in which business data are used to create the most efficient boundary of production against which others can be compared. In any improvement programme, constant reassessment is required to gauge what changes are being achieved.

It is an inherent assumption that managers can change their ways, if their desire is sufficiently

DOI: 10.1079/9781800627246.0010

strong. This relates to the argument on 'plaster or plasticity', referred to in Chapter 1, which looks at whether people are fixed in 'plaster', or whether change is possible. The research makes it clear that 'plasticity' is the reality.

The earlier chapters have covered many of the aspects that make up a manager, and thus refer to what might be changed with benefit. The following were noted about recognized experts.

- They were excellent in a limited domain (area of expertise).
- They were good at defining a problem (anything that requires a decision).
- They can accurately observe the relevant information.
- They can perceive meaningful patterns (structure of a problem).
- They have superior short- and long-term memories.
- They come up with the best solution.
- They have good self-monitoring skills (improve their expertise).
- They spend time on sorting problems they have not previously experienced, and from this create and improve intuition (tacit knowledge).

Thus, good managers will have all these attributes. In addition, it has been pointed out that managers need to:

- know the technology;
- be good at recording the relevant items they have observed;
- be good at anticipation and planning;
- have a good knowledge of decision rules and principles;
- have excellent people skills;
- have an appropriate personality, or management style; and
- have appropriate intelligence.

This is clearly a demanding list.

Intelligence wise, a mind able to think logically is important, as is one that can hold and retrieve the relevant information as required. Logical thinking is related to learning from experience, which involves self-criticism and analysis. Whatever the objectives, the manager's attitude to risk impacts on what actions are optimal, but extreme attitudes are probably not helpful. And the manager's belief in what control s/he does have (locus of control) is relevant in that it might be different from reality and therefore impede progress. Risk and control attitudes are more than likely correlated with a manager's inherent management style, or personality.

The five factors in personality impact on a manager's style, which in turn impacts on likely managerial success. Conscientiousness is obviously important, as is 'openness' for it confers original thinking and a spirit of 'trying out the new'. In order to develop good relationships, a degree of extroversion is beneficial, as is agreeableness, for this confers a benign approach to others. A certain amount of 'anxiety' is valuable in that it stimulates timely action and care, but too much causes stress and subsequent irrationality. In efforts to improve management, thought must be given to changing the expression of these personality characteristics.

It was pointed out earlier that intelligence is partly genetic and partly developed (crystallized). It does appear, therefore, that intelligence can be enhanced with appropriate training. Certainly memory can be improved as can creativity (an important aspect of coming up with solutions), logic, learning from experience, and spatial aspects such as understanding maps, perspective and all tasks requiring spatial organization. Much of the mind's work relies on pattern matching, and so developing images, patterns and their efficient retrieval is an important function of the mind.

The model of ability discussed in Chapter 3 brings together the major factors reviewed above. Besides style and intelligence, it is clear that experience is an integral part of the evolvement of a manager's ability. Any management skill improvement programme must look at how the lessons of experience can be assessed and better utilized.

Reasonable intelligence impinges on learning from experience through problem recognition and the use of benchmarks. The ability to form appropriate chunks in any problem is also important, especially in primary production with its biological connotations and consequent interrelationships, and indeed in any production system involving many and complex interrelationships. An appropriate intelligence also enables decisions which allow for the dynamic nature of production. Effectively, there are interrelationships across time which must be understood and integrated into decision making. Further management complexity comes about through risk and uncertainty, the impacts of which must be clearly understood, as, if nothing else, primary production in particular and related industries is certainly full of risks. A flexible management style is necessary as initially made plans may need changing as production unfolds. A person who cannot readily change and think 'on their feet' will find success difficult.

A reasonable IQ is also important in understanding the decision rules that lead to optimal production. For efficient resource allocation, a manager should follow the rules derived from a study of production economics, and also of similar economic theories such as cost–benefit analysis. Fortunately there are many texts devoted to such subjects, though few managers formally study them, leaving their logical skills to lead them to the lessons such as the 'marginal return equals marginal cost' rule.

Chapter 5 considered the specific competencies managers believe were important. Clearly any training programme must take this list into account. It will be recalled that the successful general competencies heading the list were:

- risk management;
- observation;
- anticipation;
- negotiation;
- planning and the associated analytical skills;
- learning from experience;
- skill in dealing with people;
- successful implementation of plans;
- an excellent knowledge of the relevant technology; and
- an ability to integrate solutions across all components.

Also regarded as being important is an ability to achieve:

- timely action following timely observation of the signs that indicate action is needed.

Furthermore, part of fulfilling the competency list is having good skills for:

- watching;
- reading; and
- listening.

The challenging lists continue.

Chapter 6 highlighted the common decision biases and noted that stress can give rise to irrational decisions, and Chapter 9 mentioned that emotional factors can, and do, influence decisions. While methods of stress alleviation were discussed, there is no simple answer to overcoming bias. In a general sense, any managerial process that gives other-than-optimal decisions is biased. Overall, then, the earlier chapters have covered the many things that make up a manager and their skill, including intuition in Chapter 7. This clearly models and summarizes the base skills required. The importance of a manager's personal characteristics was further emphasized in Chapter 8 covering 'issues in management'.

The essence of a plan to reduce decision bias and improve managerial skill is to tease out the core factors important to any one manager and institute a programme of correction for situations where deviations exist between perfection and reality. While most managers are either consciously, or unconsciously, working in this direction, it is important to consider the kit bag of tools available to assist this process.

The rest of this chapter moves in this direction by providing ideas and concepts on the processes of learning, the kinds of benefits possible from a good programme, what topics should be tackled, ideas on the resources available for helping, and, finally, a list of the training methods that might be considered. The chapter is completed with some concluding comments on the book.

Learning Processes

Everyone learns best in slightly different ways. You have already been introduced to Kolb's test which lets you conclude whether a person is a 'concrete' thinker who best learns from practical situations, or, at the other extreme, is capable of learning from abstract resources such as a textbook. Whatever the case, it is fair to note that most people learn from practical exposures, though this may not be the most efficient in that a computer screen, or book, is a cheap resource, if in fact a manager can obtain meaningful skills and knowledge this way. But few can.

First, *learning will seldom occur, if the person is not a willing student*. An open mind that is keen to change and absorb any lessons on offer is a good start. This is not that someone with opposite inclinations will not learn and change their views. Clearly, some bad mistakes that are financially disastrous will most likely lead to a change of views and improved management, provided the person can work out what went wrong and why.

Moving a manager into an appropriate frame of mind can be achieved by providing information that shows improvement is possible. This information might be comparative business statistics ranking the particular business somewhat down the list. Managers who are not overconfident are clearly going to be more receptive. Furthermore, if what is being learnt is exciting there is a better concentration on the process. Ewell (1997, p. 9) comments 'learning occurs best in a cultural and interpersonal context that supplies a great deal of enjoyable interaction and considerable levels of individual personal support'.

Second, *the manager must be exposed to improved methods, knowledge, ideas or whatever is appropriate for her/his situation*. This means pointing out the process currently used by the manager and comparing this with an improved approach. A simple example would be methods of creating a whole-business budget. In some cases, perhaps the manager has never created a budget, and so the training starts with a blank slate, though it would be surprising to find managers who have never produced some kind of rough budget, even if perhaps mentally.

Practice is then required to reinforce the correct method and ensure the process has been learnt. This step may take some time, but eventually will be achieved. Think back to learning to ride a bicycle. Initially it was a difficult and painful experience, but eventually the process became in-built and automatic no longer requiring obvious conscious thought. The procedure for solving the problem has become part of a manager's intuition.

There is no reason why this learning process should not equally apply to all the skills required, though some will take longer than others. An example of the latter might be forecasting international prices of a commodity.

Research on children (Nuthall and Alton-Lee, 1993; Nuthall, 2007) has concluded that repetition is important in that there must be a number of instances, and types, of relevant experiences for the student to develop a construct or understanding of an idea. The time interval between experiences is also important. It also seems that much of the learning occurs from within student group interactions and discussions, in contrast to the formal lessons. Most people would recognize this situation in that seldom do you comprehend a concept or idea without several exposures and reinforcement. When it comes to simply remembering a fact this is probably different. If someone tells you that it is necessary for you to pay your tax bill by, say, the end of July, most people will only need to be told once!

In a similar vein, Enos *et al.* (2003) believe 70% of learning relates to informal learning leaving 30% occurring through formal training. They comment (p. 379): 'Informal learning is a continuous cycle of challenging experiences, action and reflection. Informal learning for managers is a social process (social interaction with others in the workplace).' It is interesting that the situation seems to be identical no matter what the age of the learners.

There are many examples of this kind of conclusion. One other example (McCall *et al.*, 1988) found that 30 of the 35 managerial job skills they believed are important were learned informally, and Beaudry *et al.* (2005, p. 1) stated: 'The human resources literature suggests managerial skills are difficult to codify and learn formally, but instead tend to be learned on the job.'

All this research emphasizes the interactive and social aspects of learning; thus, sometimes learning can be assisted in the rural environment through managers getting together in groups (Fig. 10.1). These groups can be formally organized, or perhaps might just be 'neighbours' regularly contacting each other (Nuthall, 2016). Similarly the business family might be involved.

The third important factor in learning is feedback. Having absorbed a lesson and implemented it (e.g. learning about and assessing the risk associated with winter stock feeding, or perhaps the use of a new automatic machine, while trading the produce on a free market using, hopefully,

Fig. 10.1. Managers enjoy seeing the systems and procedures their colleagues use. Such 'in the field' comparisons stimulate an examination of the manager's own system, and critical inspection. (Courtesy of David Hollander.)

successful market price estimates and subsequently ensuring certain reserves), observing the resultant outcomes clearly leads to improving the decision rules created. Effectively, as Kolb *et al.* (1974) noted, 'learning is a continuous process of concrete experiences, reflective observation, abstract conceptualization, and active experimentation'. Some refer to this as 'experiential learning' (Kayes, 2002). Feedback might come from the manager's own observations and records, or perhaps from consultants and colleagues. Positive reinforcement is always important.

Mention was made in Chapter 4 about Kelly's idea of 'man the scientist' in which he is always trying to come up with constructs (rules of thumb) that suit the evidence observed. Such constructs should be constantly under review to provide better systems that match the observed environment. Clearly, learning is a continuous process, and is what leads to intuition (tacit knowledge), that mysterious thing which drives us all, hopefully, in the right direction. A manager's current knowledge and understanding will clearly influence the interpretation of outcomes and thus the lessons to be learnt. In a way, humans cannot prevent themselves from learning in that most are striving to improve, it is just that some are better equipped to achieve this than others.

Self-'critiquing' is the fourth factor. The knowledge of one's strengths and weaknesses can be included in this aspect. It has been noted earlier that an important skill is being critical of everything observed, so that any information or idea must pass the manager's internal test of reliability and accurateness before it is accepted. Similarly, being positively critical of one's analyses and constructs is important and is one of the processes involved in learning. This is part of the 'man the scientist' idea and leads to examining one's own processes leading to improved decision systems and knowledge.

Thus, in summary, a theory of learning that is widely accepted involves having a person who:

1. Is willing to change and has an open mind.
2. Can accept that improved ways do exist and is willing to search these out.
3. Can practise and re-practise a new approach until the lessons are learned (experiential learning).
4. Is given feedback, either self-created or from others.
5. Is self-critical.

As noted earlier, these are the attributes of experts and it is stressed these learning processes are how they actually become experts. Of course, other approaches do exist, such as reading a textbook, but it is clear that managers seldom use, nor benefit from, these formal approaches. The research does emphasize that managerial skills do need to be learnt on the job. If the objective was to become, say, an art historian, the situation could be quite different.

Consultant Bruce, being heavily involved in training managers, has ideas on managers' learning attributes. He is a strong believer that managers are kinaesthetic learners: 'Show me how and I can see what to do.' In a similar vein the 'trial-and-error' experiential approach is important: 'Give it a go and you will soon learn.' And Bruce comments that managers learn a lot from their peers, particularly the respected managers, proving the point that if you can get the leading managers to change many others will follow. Overall, Bruce believes that a manager must have an open mind: 'Hey, I've made a mistake, how can I learn from this?' A consultant can be an important catalyst in this learning process through being a mentor and directing the manager to role models.

Benefits of Changing

Most managers would like to improve their skills, and every manager can improve, if he or she has the desire. Clearly, the benefits involve increased profits and/or greater satisfaction with, possibly, greater resource efficiency. The amount of change will depend on the particular manager, and the resource devoted to the exercise. As with all investments, and a change programme is nothing less than an investment, the costs must be weighed against the benefits.

Everyone embarking on a management improvement project must assess their situation in this respect. Eventually, however, improvements can become almost costless in that the right attitude of observation and change can become automatic, and the manager becomes an expert with the constant review ability.

It is not possible to give figures on, say, profit increases relative to time invested, as each situation will be unique. But various comments of a general nature can be offered. Frequently, people running short courses for managers seek feedback on whether the participants believed they gained. Virtually all published data shows a positive response from the managers.

For example, Cameron and Chamala (2002) developed a number of tests which they used to test managers' views on the value of an extension programme. The results compared pre- and post-training test scores and found considerable gains from the courses.

That study was conducted in Australia. Hanson *et al.* (2002) provide a US example which showed that training in financial and production management significantly altered managers' views of various techniques with the managers recording they found the programme extremely valuable. Hopefully, the managers then went away and implemented worthwhile changes to their management. More recently Jackson-Smith *et al.* (2004) found a link between a deeper understanding of financial concepts and greater returns following managers' attendance at courses. The managers also testified that they found the courses useful. The advantage of this latter study is that not only were the managers' views assessed, but the profit outcomes were also recorded.

In a more general sense, over the years, there have been many studies on the value of education in the general population. Psacharopoulos and Patrinos (2004) provide a review of many of these studies. For example, Ashenfelter and Rouse (1998) found there was a 9% increase in wages from each year of schooling using 700 identical twins to give comparisons. Furthermore they maintained that family background explains about 60% of the variance in school attainment. In an attempt to further quantify the relationship, Bowles *et al.* (2001) found that wages were given by the equation:

$$\begin{aligned}\text{Wages} = {} & 0.196\text{years of schooling} + 0.081\text{IQ} \\ & + 0.035\text{years of experience} \\ & + 0.025\text{parental socio-economic status (SES)} \\ & - 0.096\text{number of children}\end{aligned}$$

It is interesting to note the schooling and experience figures. Of course, this is for a sample of the general population, many of which would be employees, but nevertheless, the benefits of training and experience are strongly positive. In business, it would be expected that experience would be more important.

When it comes to specific primary production and related training, Kilpatrick (1997) notes that managers with more education, and who attend more courses, have greater profitability than their peers. Kilpatrick (1999) also shows that the more the education, the greater the number of changes a manager makes each year in their management, and the greater the number of training courses attended. It is also noted that US experience shows that 50% of the productivity changes made are due to learning on the job. The 'learning on the job' leads into tacit knowledge (intuition), which Sternberg *et al.* (1995) believe helps explain job success. They also note tacit knowledge was not correlated with IQ, and IQ explains about 25% of job success.

With respect to formal courses, both schooling and vocational short courses, it is clear that the overwhelming evidence points to very positive benefits, though it must be noted that the benefit to each manager must depend on her or his unique situation. It must also be noted that for short courses, most attendees used in the research were probably managers who were enthusiastic about learning. If the managers had never attended a course before, and did not wish to attend but had somehow been persuaded to attend, the results might have been different. But you cannot say whether the outcomes would have not been good, and indeed perhaps the improvements might have been even greater?

In that considerable learning occurs 'on the job', the benefits probably relate to the help and guidance available. There is, however, no research quantifying the value of 'one-on-one' tutoring. What are available are the results from the extensive research on the benefits of personal counselling. Overwhelmingly the results are positive (e.g. Atkinson *et al.*, 2000), so you would imagine that the same would apply to similar management tutoring. Definitive research on this for managers must await the future. What can be noted, however, is that in countries where private consultants are available, they are largely fully employed, and in cases where state-funded consultants are available, they too are generally very busy. Thus, the managers must think that their help is valuable, though some of the assistance is rudimentary (such as making financial comparisons and yearly budgets) in contrast to working on basic management skills. However, not all managers make use of the consultant's or extension officer's 'one-on-one' help. If more was available, it is not clear whether they would similarly be fully occupied.

Topics and Resources Available

Managers have a wide array of resources available to help in improving their skills. Material available on the World Wide Web can be accessed by large numbers, and certainly books can be accessed by all. In many regions, both long- and short-duration courses are available on a range of topics specifically for managers. Some courses are run by state-funded extension organizations, and others are offered by non-government organizations, usually on a fee-paying basis. Some universities and other educational organizations have programmes accessible on the Internet, or by correspondence, and even attendance in person where this is practical with respect to the location of an organization.

While some managers will be very clear on what they wish to learn, others will need assistance in selecting what to study, and in starting off. Even for the former, however, assistance can be useful for selecting resources that they will find useful and appropriately positioned.

Business colleagues can help, as can consultants and extension personnel. Managers' organizations are sometimes similarly useful.

For planning skill improvement programmes, it is useful to have a list of topics that might be addressed. Planning, as the first skill, starts with the general skill of observation and understanding the facts; so, successful courses on observation (Nuthall, 2010) might be a starting point. This is followed by the general skill of anticipation and all that entails, and then any programme would not be complete without studying risk and uncertainty.

Topics within observation which need to be considered include:

- 'what' to observe;
- skills necessary for optical observation (scanning, interpretation, re-examination, review and repeat, snapshot storage, conclusion, notebook jotting);
- factors in reading (active reading, skim, summarize, question, recite and review, conclusion);
- listening skills (active approach with feedback and confirmation, attention, reception, perception, open mind, etc.);
- listing and prioritizing objectives (manager, spouse, family, other owners, etc.);
- memory development (storage, retrieval, practice of repetitive association, remarkableness, visualization, etc.);
- problem definition;
- deciding relevance (critical view of proposals, offerings, etc.); and, finally
- record keeping (what, how, legal requirements, cost and return of records, etc.).

No doubt you can think of other aspects of observation to be added to this list.

As the future is uncertain, the skill of making only decisions that must be implemented in the 'here and now' must be acquired. If made early, decisions that do not need immediate implementation may turn out to be suboptimal once the time for implementation arrives. The conditions are likely to have changed. Thus, while a manager might have in mind what needs carrying out at some future date, the skill of having 'in mind' several possible actions is important, so that once the future becomes the present the decision suitable for the current conditions can be implemented.

Of course, decisions should not be delayed beyond their 'use-by date', for, in such cases, the operation will not be 'timely' and the outcomes less than perfect. It is a fine line. A manager must learn an attitude that copes with delaying decisions to the last minute. Methods encouraging managers to have this approach are clearly important and require the manager to be 'clear headed and to hold their nerve'.

Anticipation must also include learning about:

- planning horizons (how far into the future is anticipation required to ensure the actions implemented now are optimal);
- strengths, weaknesses, opportunities and threats (SWOT) analyses to allow appropriate planning;
- developing imagination, creativity and visualization;
- ensuring practicality (realistic visions); and
- forecasting (reading, technology, markets, understanding relationships – both technical and economic, rules and regulations, etc.).

Imagination training involves:

- brainstorming;
- use of random words for starters, and such techniques;
- logical thinking; and
- learning to break from traditional concepts and watching for emotional blocks.

Visualization is effectively a mind-based experimentation and involves pattern matching, so exercises to develop these skills are necessary. These mind experiments must allow for the correlations and interrelationships of the business components; 'what impacts on what' must be a topic constantly practised.

Anticipation also involves putting plans on to paper and thus requires:

- budgeting skills (physical to financial to cash flow);
- critical path analysis;
- project management (monitoring and control, etc.);
- knowledge of all the regulations around change (employment contracts, resource

consent applications, environmental impacts, etc.);
- time management; and
- feed budgeting where animals are an important component of the business (and, similarly, for any other resource that will be in short supply or scarce).

When it comes to implementing plans, it is important for the manager to have the confidence to act on her or his convictions. This is a management style factor which needs consideration. Some managers 'dither' and action suffers, ending up in a 'muddling along' type situation. Thus, studying strategies to provide confidence and conviction is important to some (start in a small way perhaps, confer with supporters, etc.). Similarly, of course, the manager's management style in general may need attention, but this 'topic' probably requires more than just a simple study of the procedures involved. The next section considers methods of achieving a change of a manager's management style. The topic of 'people skills' also comes under this area. And the same comments apply to developing logical thinking and other components of intelligence.

Under risk management the topics that must be covered include:

- probability, chance and uncertainty;
- sources of risk (prices, weather, employees, markets, transport, etc.);
- tools to reduce risk and uncertainty (diversification, contracts, insurance, low-variability products, etc.); and
- calculations necessary to analyse alternative strategies.

One of the difficult topics in risk is understanding the idea that while you might make an optimal decision, in reality it turns out to be wrong relative to what you would have done given hindsight. Thus, the nature and features of risk and uncertainty must be studied. Similarly, determining a person's risk attitude must be covered, and how this can be used in planning.

For the resources available for each of the topics, comment has already been made about the Internet, and the availability of books. Similarly there are courses offered in many areas (e.g. one Australian group, Farm Management 500, has courses on financials, administration, risk, land, succession, business safety, marketing, machinery, staff management, people skills, crop and stock production). Books and articles abound on skills like negotiating, conflict management, assertiveness training, communication, listening and the like. Books on production economics as related to agricultural production also abound, as do books on price determination and marketing.

Finally, it is important to mention decision support systems (DSS). This refers to business computer-based packages (Fig. 10.2), and also Internet-based, that end up providing in-built training in management skills simply through their constant use. Any manager using a good package will, depending on the system, gain skills in observing and recording relevant data and information, and analysing the data to provide useful decision support. Packages should also relate to the manager's goals and thus stress this aspect of being a good manager. Artificial Intelligence systems are also likely to become more dominant as more experience is gained and experimented with.

An inquisitive manager will want to know the assumptions inherently contained in the packages and modify these to suit her/his situation where possible. Furthermore, integrated systems will stress just how the components of a business are interrelated. Use of these packages will enhance a manager's observation, anticipation and analytical skills. They should also allow for risk and uncertainty.

Training Methods

Overview of methods

Mention has already been made of how successful learning is most likely experientially based. The kinds of resources available to support any skill enhancement programme have similarly been listed. Skills that simply involve the use of listening, reading and watching can be picked up in a straightforward way by managers, either on their own, or through some course or group activity. Examples are ideas on methods of reducing

Fig. 10.2. In the early days all but the far-sighted SME managers scoffed at computers. But now they are an important part of their management team – in this case, a rural contractor checking the programme and plans.

risk and uncertainty, understanding the basics of negotiation skills, and so on. For skills that involve other than simply learning the facts, experiential-type systems become important. This is where attitudes and tacit knowledge need enhancing.

Where an individual is working by her/himself on an improvement programme, and everyone should be doing this whether or not they are involved in any formal programme, the person must learn to constantly review what has happened with a critical view and tease out the lessons as a self-learning experience. Of course, few people actually live by themselves, so sharing some of this review and analysis with others locally based is nearly always valuable.

Where it is possible, formal assistance in reviewing the lessons of experience can be sought. This might be through extended family members, and/or compatible neighbours. It also might be through a group arrangement where the group sets itself up as a group mentor-type situation. Such groups might be organized through a manager's organization, or perhaps just by locals who are keen to improve. Groups might seek professional help to guide their activities and offer ideas (Nuthall, 2012).

The other alternative is to seek 'one-on-one' professional help in making sense of the lessons to be had. The professional might be an agricultural consultant, or extension officer, or it might be someone who is skilled in decision making but does not necessarily understand agriculture.

This whole process of making use of experiential learning requires:

- noting carefully the sequence of events;
- the conditions that existed;
- the outcomes; and then
- reviewing them to see whether some other kind of response, or decision, to that taken might have been more appropriate.

Where others are involved, it is useful to retrace the manager's thought process in coming to a conclusion and action.

In non-professionally supported groups, people take it in turns to expose what they have done to encourage both self-analysis and group input. Such groups are a regular feature of urban life, but often for different reasons (business mentoring and weight-loss groups, for example). The group members might also quiz each other in turn for they may expose problem areas that the targeted manager might not even recognize as a problem. These processes must, of course, be carried out sensitively. The members of a frank and open group must be compatible and possibly self-chosen. Furthermore, an absolute rule must be to retain complete confidentiality unless otherwise specified, for without this it is doubtful whether managers will have the confidence to recount the full details of their thoughts and processes. Thus, one benefit of group member examination is creating better skills in problem recognition.

To simulate situations that can be self-, or group-, assessed, computer-based simulation games can sometimes be used. These attempt to create decision situations that allow a manager to practise the skills acquired and provide 'hindsight' information that can then be considered for providing lessons on what went wrong in the decision making. For simulation games to be useful they must be realistic and relevant to the environment under which the managers are working. There is some evidence that games do have significant benefits in student learning (e.g. Nielsen, 1974; Menz and Longworth, 1976; Curland and Fawcett, 2001).

Mentoring approaches

Whether working on a 'one-to-one' basis, or in some form of support group, there are some principles and approaches to mentoring that have been developed in the world of counselling. These proven techniques will undoubtedly work when trying to improve the attitudes and management style of managers. This is a separate problem compared with simply learning facts, figures and techniques. Generally the methods rely on:

- understanding why a problem exists; and then
- attempting to modify the current thoughts and behaviours that gave rise to the 'problem'.

The word 'problem' is used, of course, to represent any situation that needs changing. An example is where a manager tends to get over-anxious. Another is where the manager gives little thought to the future prices and the collection of information that might help a manager be better equipped to anticipate what might happen in the markets. In this example, there is usually plenty of help from the marketing experts who often write in popular journals. But the view of the manager about what s/he should consider may need altering, and the manager certainly has to look at the reports from an informed perspective to enable assessing their importance to her/his business situation.

Management mentoring can be defined as a relationship between two or more people designed to improve the management skill of a manager. For success such a relationship should be:

- non-judgemental;
- open and frank (for obvious reasons); and
- involve trust, sympathy and understanding.

The objective is to provide an environment and approach which ensures everyone is relaxed and confident, so that they can express their thoughts and ideas without any form of criticism.

Bruce the consultant is a strong believer in mentoring arrangements, but he similarly expresses the importance of the need to 'trust one another', and the need for 'accountability between each member of a group'. He also believes that a mentor group can find it very valuable to visit other regions and systems to get them asking 'why does that manager do it that way?' This leads to the managers questioning the reasons they follow in their particular systems, and perhaps together concluding that a change might be valuable.

Furthermore, *learning about the background to a problem* clearly helps others suggest alternative approaches that might solve the problem. An example might be where a manager tends to be very short-sighted when planning and making decisions. Perhaps s/he learnt her/his basic business skills in a very reliable set of conditions (climate, markets ...) and needs to understand more about the current environment. Another example might be a manager who has poor negotiation skills. Perhaps s/he is simply a strong introvert and is fearful of dealing firmly with people suggesting alternative solutions to whatever is being negotiated. Once the manager understands her/his situation better, perhaps s/he and the helpers can develop strategies to overcome this situation. Maybe the manager should never agree to anything straight off and then consult others before suggesting a change to the proposed deal. The objective is to give the manager insight into why s/he has always fared poorly in negotiations, and thus develop an awareness of how negotiating might be better tilted in her/his favour.

Seldom is it possible to alter a manager's attitude or approach overnight. It will normally be necessary *to return to the 'problem' for reviews from time to time to constantly assess what progress* has been made and reinforce the solution offered. Reinforcement is often useful in the form of some reward or other. Perhaps the manager can reward themself with a half-day fishing, or golfing, if the mentor deems that progress has been made.

Another way to explore and reinforce appropriate management is to visit and *discuss with people who are good at the skill under consideration*. Managers generally enjoy talking to other managers about their methods, and so where a manager is generally regarded as an expert, for example, in dealing with risk, then their skills need tapping into. This method is called 'modelling', you are modelling off an expert manager.

Then what is known as *role playing* can be useful. This is about practising an appropriate management skill. Perhaps a good manager can act out how s/he deals with a conflict situation so that another manager who often has difficulty with staff might then practise dealing with a difficult situation. One manager acts out being an employee, and the other acts out being the manager. Of course, before doing this they must have sorted out what they believe to be the methods that should be used in conflict management. The manager recognized as being an expert is a starting resource, as are books on the subject. Where a 'one-on-one' situation exists, the consultant will have at least the theory of conflict management and so can role play the situation. Effectively, this approach is all about practising in a supporting and understanding environment. *Assertiveness training* might follow the same approach where a manager needs to learn to be assertive, but not aggressive. Others might call it 'sticking to your guns'.

What is known as *desensitization* can sometimes be of benefit. This relates to situations that are of real concern to a manager to the extent that s/he does not function efficiently. While this might be called stress, it usually refers to a particular situation causing concern, whereas for all other situations the manager is totally relaxed. An example might be bidding at an auction for produce to sell. Maybe in the past the manager has made some bad decisions to the point where s/he now finds it difficult to operate rationally at auctions for fear of getting it wrong. The desensitization approach attempts to replace the concern with something incompatible with the anxiety. Perhaps simple physical relaxation will solve the problem. The manager starts by imagining being at an auction and at the same time practises muscle relaxation starting with the foot muscles and moving up the body. In the first instance, a comfortable chair should be used! The next step might be some role play, then eventually actually going to an auction that involves something minor and then slowly moving up to an important situation. This desensitization approach is called 'extinction'. You are trying to replace the problem with an alternative feeling so that the problem is extinguished.

While working with others on perceived management problems is one option, there is nothing to stop a *manager from working on her/his own*. Indeed, it would be surprising if most managers did not, it is just that they may not be very effective. *For self-help, it is important to learn to monitor your own behaviour, and to use self-reinforcement* (maybe that half-day at the golf course, or just sitting in the sun with a book!). Initially the manager must recognize the problem/s. This is where benchmarks come into play, and also keeping records of what happens, and noting your approaches to various problems. Often just the exercise of keeping good records starts the manager thinking about the situation and consequently considering ways to improve. Perhaps, for example, a review of a manager's diaries makes it clear s/he is reviewing stock levels on a regular weekly basis. It is easier to stick to a simple routine rather than keep better records allowing a constantly updated figure reflecting the constant market movements. Having noted this habit, maybe the manager will then consider whether an alternative approach would be better.

Usually a manager will need to practise *taking note of her/his feelings and processes*. This is where self-honesty and openness are important, something only the manager will know occurs, though sometimes a spouse can help in the process from their knowledge of a person gained over many years of observation. Effectively a manager needs to conduct a personal SWOT analysis:

- strengths;
- weaknesses;
- opportunities; and
- threats.

The opportunities relate to training and improvement opportunities, and the threats to distracting situations. Sorting strengths and weaknesses is where the main effort must go.

And *with self-help, it is possible a mentor can help develop this skill* through being a simple facilitator encouraging self-analysis and efforts to change thoughts and processes. A mentor in this role must be totally non-judgemental and empathetic as well as encouraging the person to come up with his or her own suggestions through supportive prompting. Of all approaches, this one requires the manager to be convinced that self-analysis and change are highly desirable. They need to have considerable self-control, or to develop it.

At the other end of the scale, *group approaches are particularly useful where a manager has problems relating to other people*. The manager can see the reaction of others as her/his ideas and comments are expressed, particularly in role-play situations. Groups are also a good approach where there are a number of people with the same 'problem'. For example, for managers who have an inappropriately low 'locus of control', sharing their views and ideas for improvement will be a shared and reinforcing approach. To initiate this, of course, someone has to conduct a test and organize a meeting. Perhaps a professional consultant might do this for a subsection of the managers who have contracted him to help.

In summary, to change attitudes and inappropriate tacit approaches, systems must be set up that provide interpersonal relationships that exhibit trust and understanding. The system must then provide reassurance and support, bestowing a belief that positive change is indeed possible and real. It must also provide an avenue by which problems that worry people can be shared, making them seem less serious and damaging. On the other side of the coin, the process should provide reinforcement for the management skills that are currently good, and similarly of improvement as it evolves.

Providing an understanding of the source of any incorrect management skill and procedure is also usually beneficial. The knowledge that change is possible is also obviously extremely valuable, and there is certainly evidence that managers improve given the desire and right support. Visualizing what might be possible is always a way to set the goal posts, and records reviewing where a manager is in the stage of reaching the 'goal posts' are part of the necessary feedback (Aspinwall and Taylor, 1997; Taylor *et al.*, 1998).

Where *mentoring takes place, this is often referred to as a 'participatory' approach* as the manager is participating in contrast to being lectured at. Some Dutch evidence in which managers were mentored in small groups shows just how effective this technique can be. Wolf *et al.* (2004, p. 205) conclude: 'The identification of weaknesses is an essential element to motivate managers to improve their entrepreneurial qualities' and '[t]he group learning process contributes to the improvement of individual entrepreneurship, facilitating managers to learn from colleagues and to become more conscious about personal strengths and weaknesses. The comparison of individual results and the discussion with colleagues are very useful methods to create a group learning process' (p. 206).

Case managers' views

As he frequently pointed out, Margrave always enormously enjoyed trying out new technology and ventures. This adventurousness fed off professional agriculturists. Margrave discovered two particularly compatible consultants, one employed by a government-funded ministry and the other in private practice. These people became his mentors with whom he shared most aspects of his operation, and they in turn came to him for comments on ideas and technologies they were keen on. This was a mutually beneficial arrangement which gave Margrave considerable confidence, assistance and support. One suspects that his high intellect and forward-looking attitude went beyond many managers' and so he bypassed nearby colleagues to provide the kind of mentoring he looked for and found it in the professionals who happened to live and work in his area. Margrave comments: 'It was a two-way relationship with all of us bouncing off each other with all sorts of ideas.'

Margrave believes intuition is an important aspect to his decision making and he does have considerable confidence in the decisions that seem to just appear. He comments: 'I'm quite an intuitive manager. I guess this arises from the intelligent use of experience. I've been doing all that kind of thing since I was a little fellow so I must have learnt something.' To Margrave, reflection on all his experiences to come to a conclusion on the correct approach is important and adds to the development of reliable and correct intuition. Careful attention to appropriate analyses over the years helps develop this intuition, as does a healthy review of the current conditions to ensure that the intuitive response is really applicable. His story of the potential holiday home comes to mind; his strong intuition led him to a logical conclusion.

When asked if he thought his management style could be changed for the better, Margrave was quick with 'probably not greatly', but on reflection he was not so sure. He ventured it was possible to 'change round the margins' and noted: 'I gained in confidence with encouragement and support and consequently improved. If feeling confident you make the right decisions.'

Margrave also found mentor groups valuable and obtained valuable thoughts on current issues. He noted they were 'a great sounding board, a great medium for all sorts of issues. We trusted each other and treated the discussions with total confidence. The discussions did not distinguish between business and family and so our total lives were under scrutiny.' Margrave also found his involvement with a national group designed to explore all issues facing his industry valuable from a technical viewpoint. These meetings invited leading scientists and practitioners to present their ideas and engender critical discussions and eventual conclusions. But this forum of managers was largely divorced from each other's immediate problems.

But what was intimately involved in Margrave's decision problems was his use of computer software to analyse his data. 'My consultant started me on using the computer, and I learnt reasonably quickly.' He admitted he started by playing computer games, but 'now I have all the skills and get totally bored with them ... can't be bothered. I need something more challenging and have found this in sorting out business issues.' Margrave is involved in planning all the feed use on the business and constantly updates his estimates and calculations forecasting supply and demand. He also has computerized business maps for planning activities and resource use. Spreadsheets and word processing are now second nature. But the important thing for Margrave is that 'half the benefit of using the software is the need to process the information which then passes through my mind ... this active involvement starts the subconscious going and stimulates the intuition ... the answers become obvious.'

Hank also finds his computer a very important part of his management 'team' with its constant use on cash flow updates, annual budgets, input planning as well as its use to access the dairy industry central records. These provide monthly updates on output as well as the ongoing totals. Another important, but passive, member of his team is his diary. It gets used to record goals, job lists and achievements, leading to continual job list updates, and their re-prioritization. The diary gets attended to the first thing each day. 'I get muddled if I don't write it all down, and I certainly don't like feeling out of control.' Given the size of Hank's operation one can understand how he feels. The diary provides order, and its use is a chance to think through all the issues.

Hank has made, and does make, constant use of mentors and groups. He used to belong to a reasonably sized discussion group, but comments: 'I found this group useful when I first started, but now find in such a big group too much time is wasted discussing practical issues like how much ingredient was in specific silos. I now find I want more specific information relating to my situation. The meetings ended up being more of a social outing, which is, of course, very important to new people in the area.'

What Hank now finds more relevant and useful is a small group of friends who regularly review each other's actions and decisions. Hank also finds his banker is a great sounding board with whom he constantly talks, especially after sending updated budgets (which occurs on a regular basis). These relationships depend very much on the respect each has for the other. Hank's accountant similarly contributes, particularly over appropriate business structures and related taxation matters.

Hank is also a great believer in using professional consultants as a lynchpin mentor. 'I use consultants all the time for they know my situation, and also know 'me'. The first consultant I had was just great for the basic stuff such as feeding systems and calculations, but I improved my skills here, and no longer needed this kind of support. I have now moved on to a different kind of consultant with an emphasis on the business side of my operations. The trouble is I'm a strong character and consequently require someone who will stand up to me and provide challenges. Together we get it mainly right.'

Hank believes some consultants just say what the manager wants to hear, and tend to be 'pollinators', simply passing from one manager to the other the best ideas: 'This is a waste of time for experienced managers.' He concludes: 'The personality mix is very important, and the guy has to have credibility in the eyes of the manager client.'

Professional consultants' views

Bruce has strong views on the role of consultants in helping managers improve their management. Above all Bruce believes he should challenge his manager clients, so that they can assess their own situation and come to conclusions on how to improve. Of course this 'challenge' is given structure by the consultant with a view to lead the manager along logical paths.

Bruce also notes that the consultant himself must always be learning, and the methods are similar to what a manager should be doing. The consultant must always explore new situations and gain experience accordingly as well as constantly reading and investigating new ideas. Part of the learning is having extensive networks where others can challenge and provide new insights. As there is always a danger that a consultant can become isolated and stagnated, it is important to belong to professional organizations and societies.

Some consultants call themselves 'business coaches' and encourage the managers to talk about their plans and programmes, so that they can be challenged over what they are doing. This process helps the manager develop their capacity to acquire improved skills. They learn to:

- identify;
- observe;
- assess alternatives; and then
- decide and review.

In the end, consultants, Bruce believes, must be positive people: 'Here are the opportunities, let's try this one, it is very exciting.' Some managers improve with this support and the consultant can 'let them operate on their own, but others always need their hands held'. If a consultant 'puts enough fires under a client the spark to start up something new may be created'. The consultant must convince the manager new approaches are possible, and in the process share some of the risk and provide confidence. Sometimes it is necessary to 'show a manager that he can't afford not to change' and this can change his risk attitude, 'you are going to regret not doing this'.

One of the advantages a consultant can bring is a new perspective from an outsider's view, especially if the consultant has extensive networks. 'Consultants play a big role in being agents for change.' Bruce comments that managers can become very introspective with the lack of outside stimulation. He also notes that the relationship between the consultant and the manager is critical. A trust and respect must be developed, and this will only occur if indeed this trust and respect is deserved.

Any consultant must really listen to the manager and determine her/his beliefs and processes. In this respect, it is important to match the manager and the consultant personality wise: 'some match ups just do not work' (Hank also stressed this point). Sometimes initially the consultant must carefully guide the manager, but for real improvement this relationship must progress. The consultant cannot take all the risk for the manager, and, besides, doing this can be very risky for the consultant himself. Success is dependent on the outcome of many risk factors over which the consultant has no more control than the manager.

As has been stressed, one of the first steps is for the manager to be very clear on her/his objectives. The consultant, Bruce believes, can make a real contribution in this regard, challenging the manager's professed objectives and goals. Bruce comments: 'It often takes time to drill down to the real goals.' By way of an example, Bruce noted that the professed maximum profit motive often stems from a desire to be better than the neighbouring businesses. The consultant needs to work away at sorting out these real motives, and where they originate.

Prof has similar views about the manager's objectives, and comments: 'It is absolutely no use asking a manager what his objectives are. He'll just trot out what you want to hear.' A consultant must watch, listen and observe to really find out the real objectives. The consultant must work out what the manager is doing and why through observing incidences and events.

In a similar vein to Bruce, Prof notes that managers 'need measures of success ... it's all about how you feel about yourself'. Thus, 'profit maximization is only important to the extent it is a measure of success' with managers often having quite basic objectives for which they strive. Some managers can be called 'developers' and their objective is to enhance and grow their assets, and others are simply 'managers' trying to get high levels of production which may, or may not, lead to profit maximization.

In improving a manager's ability, Prof believes you must first understand their personality, skills and weaknesses. Most would agree with this approach. Another important factor is understanding just what it is the manager wants from a consultant, as, after all, the manager is paying the bill and should 'call the shots'. One survey of a small group of consultants revealed that 37% of the managers wanted help with technical factors (how, what, etc.), 20% wanted business coaching and motivation (help with might be called the 'soft skills'), and 43% wanted a mixture of technical and coaching help. These different demands mean there are 'horses for courses', so care must be taken to ensure that the client and consultant combination works.

Prof also believes most managers learn by 'seeing and doing', so the ability to see what other managers are doing is very important. Often a manager will try something in a small way as part of 'seeing and doing'. As noted earlier, Prof commented that the ability to find out what other managers are doing is very important and in this sense a consultant can foster this 'across-the-fence' observation. Prof stressed that some managers have useful strategies to get on to other businesses. One manager who sold a particular genetic line always visited his clients each year ostensibly to see how the genetics were helping the client, but in practice it was to spend time on other businesses to see what the manager was doing, and why. Another manager was known by his children as 'side tracker' for whenever they went on a holiday, they shot off on to side roads so that dad could see and talk to other managers.

Prof considers role models are very important. Thus, manager competitions play an important role as the successful managers are available for other managers to model themselves on. In many modern rural-based industries, these competitions abound: the best manager of the year, the best financial businessman of the year, the best young manager of the year, and so on. Field days also play a part in that the winners are usually required to open their business and systems up to others.

Prof also related that one of his students studied the features of successful managers. Two important attributes were the time spent looking over the 'fence' at other managers, including attending open days on the competition winner's business, and the time spent reading each day. These are both means of obtaining ideas. A consultant can encourage and stimulate these activities including helping the manager critique what he sees and reads about.

'Leading managers are very important to a community.' Prof believes in the trickle-down theory which considers that improved systems and methods will filter through the business population from the leading managers. If the community of managers is open and considerable sharing occurs, this can indeed operate. Thus, the importance of role models talked about earlier. But this approach largely involves practical business systems in contrast to changing personal attributes. This is where an individual manager approach is necessary. Prof reckons, however, it is up to the manager to change, for you can make a manager aware of their strengths and weaknesses, but you cannot impose improvement programmes. Some managers have the basic attributes, but others just cannot be made into, for example, an entrepreneur. And when talking about families, Prof commented consultants must be cautious. They are not trained counsellors, and besides any move beyond the manager himself must be stimulated by the manager – the client is the 'king' and should control what happens.

The final word Prof had about change techniques concerned discussion groups. One of his first jobs as an adviser was organizing these groups in which managers were brought together to discuss the topics of the day, and often to inspect each other's business leading to discussions on what could be changed. These groups are extremely valuable but must be reinvigorated from time to time as they have their own natural life cycle leading to stagnation. The topics dry up, and the managers gain little more having already acquired the skills on offer. The leader of the group is critical to its success and must have an extrovert tendency to always create challenge and excitement. And to be successful, the members of any group must have the same learning needs, otherwise they get bored and drift away.

Prof's contribution to this discussion provides a different perspective emphasizing industry approaches rather than individual manager-focused systems. From a manager organizational view, and a government view, these are valuable insights.

Gender Issues

Any manager, or indeed anyone else interacting and working within an SME, should be aware of issues related to gender. If a gender issue is not handled correctly it may well impact on the cohesiveness, productivity and/or profit of the business. Similarly, for any other human attribute where employees, and others involved in the business, exhibit approaches and attitudes different from the dominant behaviour, the situation may also need special allowances. Such differences may require a manager to set in place procedures and approaches, which may involve a range of options for some situations, acceptable to all if a happy and productive workplace is to eventuate.

As gender is likely to be the most common source of differences, this section outlines some of the researched aspects and results that may impact on efficiency. In the case of other behavioural differences it may be beneficial for a manager to search for studies outlining the possible problems and solutions.

It is only in the more recent times that the issues of gender have come to be researched and considered, with thought being given to how best to develop systems and approaches that provide women with more like equal opportunities to men than in the past. Working to ensure women have equal opportunities can only add to efficiency and productivity. In that employers often seek people with experience, in the case of female applicants they are less likely to have considerable experience given the limited number of positions available in the past. Slowly this will change particularly where female applicants make it very clear in their application the special attributes they can bring to a position that their male counterparts are unlikely to offer.

Maria Minniti (2009) has written about some of the gender issues, noting in the abstract to her article:

> Women entrepreneurship presents several distinctive characteristics that differentiate it from men entrepreneurship. But variations exist also across women entrepreneurs in various countries, and between women who are involved in entrepreneurship ... Overall, the explanation for the behavior of women entrepreneurs and its distinctiveness is complex and multifaceted. Evidence to date suggests that reasons contributing to explaining these differences include demographic and socio-economic variables, subjective perceptions, and cultural factors and institutions ... Studying female entrepreneurship ... sheds light on the linkages between entrepreneurship and wealth creation, employment choices and cognition, human capital accumulation and labor market structure, employment choice and family dynamics, business creation and peace, and many others.

Due to the obvious complexities, the discussion that follows only covers a limited range of examples of the issues, and their conclusions.

Any manager must consider the particular situation they are faced with in the workplace, and community, and come up with the most appropriate solution providing the associated leadership. These approaches are likely to differ in different countries and ethnicity situations as well as for the particular type of business and its staff.

The situations faced will differ for female managers relative to their male counterparts, and certainly it is clear from the research that female managers do face hurdles different from males that will test their skills. Some examples include that female managers may not have an equal chance of obtaining government contracts relative to their male counterparts (Puntaier *et al.*, 2022), but this does vary with the type of products and services. It also seems that the complexity of the application process has an influence. On the other hand, when seeking finance female managers appear to have a greater chance of success if the lending officer is female (Blanco-Oliver *et al.*, 2021). In general, however, female managers find it harder to obtain the finance they seek (Kwong *et al.*, 2012). Yet evidence suggests female lead entrepreneurs tend to positively moderate a firm's human and financial capital and high growth (Devine *et al.*, 2019).

Beside this example, gender research has unearthed many more facets of the impacts of female managers. Some further examples are listed below.

- Females tend to be more conservative. For example, SME boards with female members tend to have lower debt according to Mínguez-Vera and Martin (2011). Furthermore, mixed boards (gender, ethnicity) tend to provide better outcomes compared with boards with less diversity. In addition, large boards tend not to perform as well as those with fewer members, and those with diversity tend to counter the narrow views of a manager (Puntaier *et al.*, 2022).
- Women tend to be less inclined to start up new ventures due to their attitude to risk being more conservative than their male counterparts (Dawson and Henley, 2015). Similarly, they are inclined to be less interested in 'entrepreneurship' (Santos *et al.*, 2016).

- Work teams with women tend to work together better than the fully male counterparts. Overall, women tend to be more cooperative team members and often mediate between women and men members. They also identify strongly with their team (Niler *et al.*, 2020).
- In larger SMEs women tend not to rise up to higher positions to the same extent as males (Hite and McDonald, 1995), thus businesses miss out on the special contributions women can provide. For example, women tend to give greater weight to ethics and social responsibilities (Ahmad and Seet, 2010) as shown by comparisons when an SME has a female manager.
- Women managers tend to more extensively use overdrafts for finance compared with male-controlled businesses, female-dominated businesses tend to be smaller than the male counterparts, and they tend to start out with less capital backing. But in either case the spouse of the manager tends to help considerably and there are more start-ups where the partner of the instigator is in a full-time job providing a security backup (Rosa *et al.*, 1992).
- Where a business is controlled by romantic partners tasks do not tend to be aligned with traditional responsibilities, though males tended to claim leadership other than in the USA where leadership was treated equally. However, for roles that could be classed as domestic it was mainly the women who took responsibility (Jurik *et al.*, 2019).
- Curiously, potential employees are attracted to positions if the employer is physically attractive (Piva and Rovelli, 2023). And women tend to be more determined, tend to better achieve goals relative to their male colleagues (Yang *et al.*, 2021) and generally 'make things happen'. On the other hand, women tend to experience more work–family conflict that may cause problems (Brieger *et al.*, 2023).
- In one study females involved in succession appeared to be concerned about gender issues relative to males. This revolves around belief in female leadership as well as trust issues (Henry *et al.*, 2013).
- Problems faced by female managers will vary from country to country. In one country, Singh and Belwal (2008) found that women had 'problems in finding the markets and distribution networks; limited opportunities for promotion and participation; limited amount of government and institutional support; absence of technological know-how and integration mechanism; and rampant corruption in an undisguised or disguised form, as major bottlenecks.'
- Strawser *et al.* (2021) comment that some economies are missing out on potential benefits where women are not given equal opportunities to men. They discuss some of the methods that might be used such as making special provision for financing women's ventures and training entrepreneurship and its components. With financing it is believed women tend to reinvest the proceeds, thus creating further growth.

While some investors might believe women involved in a business can be detrimental to an SME, there are many situations where they will bring positive contributions. One piece of research clearly shows female managers are often more effective than males in decision making and employee relationships. This example comes from a study of real estate and construction businesses (Dwiri and Okatan, 2021). Other positive research shows businesses owned jointly by women and men tend to perform better than businesses owned by men only (Simo Kengne, 2016). In another study of some 2000 businesses it was found profitability and growth were no different no matter the gender of the main protagonist (Johnsen and McMahon, 2005).

Overall it is clear every manager should have a good understanding of gender issues for they influence options, conditions and experiences, and similarly it is not uncommon for legal, political, economic and cultural set-ups to also be influenced by gender, not to mention kinship systems. Minniti (2009) emphasizes that 50% of the population are women yet the proportion of female managers is much less. It is concluded that much more study is required to discover more of the causal factors to ensure moving towards making the opportunities equitable. It is suggested research on demographic and socio-economic variables, subjective perceptions, and cultural factors and institutions should be intensified as the differences have significant implications at the macro-economic level. Other factors commented on include shedding light on the linkages between entrepreneurship and wealth creation, employment choices and cognition, human capital accumulation and labour market structure, employment choice and family dynamics, business creation and peace, and many others.

Concluding Comments Over the 'Human Factor' in SME Management

Success depends on an individual's abilities, though good luck in this risky and uncertain world helps. However, as it is not possible for an individual to make major changes to the environment in which s/he works, helping a person achieve success involves helping them improve their skills best suited to the environment. Achieving this requires understanding the person, and understanding how managers acquire the level of skills they currently have in addition to understanding how changes can be made. Changing simple factual knowledge is relatively straightforward, but changing skills stemming from inner human workings is more difficult. This last chapter has considered approaches to this problem, whereas previous chapters have provided information on the factors giving rise to a manager's skill level, and on the skills that a manager must be good at to succeed.

In improving managerial skill it is important for a manager, or her/his professional helpers, to have some kind of plan. To do this, an assessment of the types of skills that need improving must be made leading to a list of skills and processes to be used in achieving improvement. This list should be put into priority order for it is easier to work on one skill at a time. As the plan moves on, improved skills must be constantly practised and reinforced at the same time as starting on improvement of a new skill. Invariably knowledge and understanding deepen with further practice as seldom does one session create an expert.

To assess which skills need improving any tests available can be used (and several have been supplied in this book), and, similarly, the observations of a professional who has worked with the manager for some time are helpful. The manager's family may also have insights into what is required provided the manager is willing to agree to their involvement. Similarly employees and contractors may have useful ideas, again provided the manager is willing to receive what may be seen as criticisms.

In deciding on what aspects of management to improve, the manager her/himself may well have ideas that should also be included in the list. For the mentor/s, it is helpful to inspect the manager's business, and the records and accounts, to better judge the nature of the decision problems. The next step is to start the improvement process using the approaches outlined. The meetings and discussions then continue until it is deemed that the list of problems has been sorted, and the lessons absorbed and understood. The manager her/himself may well be a good judge of this.

One objective is to improve the manager's intuition, or tacit knowledge. Decisions and methods then become automatic and do not require careful and time-consuming analysis. However, the tacit knowledge must be continually assessed, and updated if the conditions change sufficiently. An example might be where there is a fundamental shift in the markets for a group of products, or perhaps the tax laws change markedly, or there might be a significant technological change.

Clearly, many managers who could benefit from skill improvement do not make efforts in this direction. Where their financial situation provides an adequate living, in contrast to an appropriate return on the investment, they may well be content. Others may wish to improve, but are fearful of exposing their situation to colleagues and professionals. In each of these cases, it is only possible to point out what is possible, and what they might be missing out on, in the efficiency stakes. If they have contact with cases where improvement has occurred, and the person makes the benefits clear, that will encourage participation. The old saying is 'you can take the horse to water, but you can't make it drink'. So true.

Bibliography

Agrawal, R.C. and Heady, E.O. (1972) *Operations Research Methods for Agricultural Decisions*. Iowa State University Press, Ames, Iowa.

Ahmad, N. and Seet, P. (2010) Gender variations in ethical and socially responsible considerations among SME entrepreneurs in Malaysia. *International Journal of Business and Society* 11(1), 77–88.

Ajzen, I. (1991) The theory of planned behavior. *Organizational Behavior and Human Decision Processes* 50, 179–211.

Ajzen, I. and Fishbein, M. (1980) *Understanding Attitudes and Predicting Social Behavior*. Prentice Hall, Englewood Cliffs, New Jersey.

Anderson, J., Dill, J. and Hardaker, B. (1977) *Agricultural Decision Analysis*. Iowa State University Press, Ames, Iowa.

Asante, E. and Affum-Osei, E. (2019) Entrepreneurship as a career choice: the impact of locus of control on aspiring entrepreneurs' opportunity recognition. *Journal of Business Research* 98, 227–235.

Ashenfelter, O. and Rouse, C. (1998) Income, schooling, and ability: Evidence from a new sample of identical twins. *The Quarterly Journal of Economics* 113(1), 253–284.

Aspinwall, L.G. and Taylor, S.E. (1997) A stitch in time: Self-regulation and procedure coping. *Psychological Bulletin* 121(3), 417–436.

Atkinson, R.L., Atkinson, R.C., Smith, E.E., Bem, D.J. and Nolen-Hoeksema, S. (2000) *Hilgard's Introduction to Psychology*. Harcourt Brace, Fort Worth, Texas.

Austin, E.J., Willock, J., Deary, I.J., Gibson, G.J., Dent, J.B. *et al.* (1998) Empirical models of farmer behaviour using psychological, social, and economic variables. Part 1: Linear modelling. *Agricultural Systems* 58, 203–224.

Austin, E.J., Deary, I.J. and Willock, J. (2001) Personality and intelligence as predictors of economic behaviour in Scottish farmers. *European Journal of Psychology* 15, S124–S137.

Beaudry, P. and Francois, P. (2005) Managerial Skills and the Theory of Economic Development. NBER Working Paper. Department of Economics, University of British Columbia, Vancouver, Canada.

Beckmann, M., Garkisch, M. and Zeyen, A. (2023) Together we are strong? A systematic literature review on how SMEs use relation-based collaboration to operate in rural areas. *Journal of Small Business and Entrepreneurship* 35(4), 515–549. DOI: 10.1080/08276331.2021.1874605.

Beedell, J. and Rehman, T. (2000) Using social-psychology models to understand farmers' conservation behaviour. *Journal of Rural Studies* 16, 117–127.

Bentler, P.M. and Newcomb, M.D. (1978) Longitudinal study of marital success and failure. *Journal of Consulting and Clinical Psychology* 46(5), 1053–1070.

Blanco-Oliver, A., Reguera-Alvarado, N. and Veronesi, G. (2021) Credit risk in the microfinance industry: The role of gender affinity. *Journal of Small Business Management* 59(2), 280–311. DOI: 10.1080/00472778.2020.1844487.

Bokemeier, J. and Garkovich, L. (1987) Assessing the influence of farm women's self-identity on task allocation and decision making. *Rural Sociology* 52(1), 13–36.

Bowen, N.K. and Guo, S. (2012) *Structural Equation Modeling*. Oxford University Press, New York.

Bowles, S., Gintis, H. and Osborne, M. (2001) The determinants of earnings: A behavioral approach. *Journal of Economic Literature* 39, 1137–1176.

Brick, C. and Lewis, G.J. (2014) Unearthing the 'Green' personality: Core traits predict environmentally friendly behavior. *Environment and Behavior* 48(5), 635–658.

Brieger, S.A., Sonbol, D. and De Clercq, D. (2023) Gender differences in entrepreneurs' work–family conflict and well-being during COVID-19: Moderating effects of gender-egalitarian contexts. *Journal of Small Business Management* 62(5), 2322–2363. DOI: 10.1080/00472778.2023.2235755.

Cameron, D. and Chamala, S. (2002) Measuring impacts of an holistic farm business management training program. In: Proceedings of the 13th International IFMA Congress of Farm Management, Arnhem, The Netherlands. Available at: https://www.ifma.network/congresses/proceedings/measuring-impacts-of-an-holistic-farm-business-management-training-program/ (accessed 29 August 2024).

Case, H.C.M. and Johnson, P.E. (1953) *Principles of Farm Management*. J.B. Lippincott, Chicago, Illinois.

Cattell, R.B., Eber, H.W. and Tatsuoka, M.M. (1970) *Handbook for the Sixteen Personality Factor Questionnaire*. Institute for Personality and Ability Testing, Champaign, Illinois.

Caughlin, J.P., Huston, T.L. and Houts, R.M. (2000) How does personality matter in marriage? An examination of trait anxiety, interpersonal negativity, and marital satisfaction. *Journal of Personality and Social Psychology* 78(2), 326–336.

Charness, G., Garcia, T., Offerman, T. and Villeval, M.C. (2020) Do measures of risk attitude in the laboratory predict behavior under risk in and outside of the laboratory? *Journal of Risk and Uncertainty* 60, 99–123. DOI: 10.1007/s11166-020-09325-6.

Chaston, I. (2009) Entrepreneurs, intuition, and small-business performance. *Journal of CENTRUM Cathedra* 2(1), 37–45.

Conard, M.A. and Matthews, R.A. (2008) Modelling the stress process: Personality eclipses dysfunctional cognition and workload in predicting stress. *Personality and Individual Differences* 44, 171–181.

Costa, P.T. and McCrae, R.R. (1992) *NEO-PI-R Professional Manual*. Psychological Assessment Resources, Odessa, Florida.

Crase, L. and Mayberry, D. (2003) Personality, objectives, attitudes, intentions and behaviour: Modelling the adoption of current recommended practice by farmers in the NSW Murray catchment. Paper presented at the 47th Annual Conference of the Australian Agricultural and Resource Economics Society, Freemantle, 12–14 February 2003.

Croskerry, P., Singhal, G. and Mamede, S. (2013) Cognitive debiasing 1: Origins of bias and the theory of debiasing. *BMJ Quality and Safety* 22, ii58–ii64.

Curland, S.R. and Fawcett, S.L. (2001) Using simulation and gaming to develop financial skills in undergraduates. *International Journal of Contemporary Hospitality Management* 13(3), 116–121.

Dane, E. and Pratt, M. (2007) Exploring intuition and its role in managerial decision making. *Academy of Management Review* 32, 35–54.

Dane, E. and Pratt, M. (2009) Conceptualizing and measuring intuition: A review of recent trends. *International Review of International and Organizational Psychology* 24, 1–40.

Dawson, C. and Henley, A. (2015) Gender, risk, and venture creation intentions. *Journal of Small Business Management* 53(2), 501–515. DOI: 10.1111/jsbm.12080.

Day, J., Reynolds, P. and Lancaster, G. (2006) Entrepreneurship and the small to medium-sized enterprise: A divergent/convergent paradox in thinking patterns between advisers and SME owner-managers. *Management Decision* 44(5), 581–597.

de Lauwere, C.C. (2005) The role of agricultural entrepreneurship in Dutch agriculture today. *Agricultural Economics* 33(2), 229–238.

Devine, R.A., Molina-Sieiro, G., Holmes, R.M. Jr. and Terjesen, S.A. (2019) Female-led high-growth: Examining the role of human and financial resource management. *Journal of Small Business Management* 57(1), 81–109. DOI: 10.1111/jsbm.12492.

Dhungana, B.R. (2000) *Measuring and explaining economic inefficiency of Nepalese rice farmers*. Master's thesis, Lincoln University, Christchurch, New Zealand. DOI: 10.21275/SR21126183926.

Dhungana, B.R., Nuthall, P.L. and Nartea, G.V. (2004) Measuring the economic inefficiency of Nepalese rice farms using data envelopment analysis. *Australian Journal of Agricultural and Resource Economics* 48, 347–369.

Dwiri, B. and Okatan, K. (2021) The impact of gender on leadership styles and leadership effectiveness. *International Journal of Science and Research (IJSR)* 10(1), 1419–1434.

Ellinger, E. and Ellinger, A. (2014) Leveraging human resource development expertise to improve supply chain managers' skills and competencies. *European Journal of Training and Development* 38(1/2), 118–135.

Enos, M.D., Kehrhahn, M.T. and Bell, A. (2003) Informal learning and the transfer of learning: How managers develop proficiency. *Human Resource Development Quarterly* 14(4), 369–387.

Eraut, M. (2000) Non-formal learning and tacit knowledge in professional work. *British Journal of Educational Psychology* 70, 113–136.

Ewell, P.T. (1997) *Organizing for Learning: A Point of Entry*. Available at: https://intime.uni.edu/organizing-learning-point-entry (accessed 29 August 2024).

Farrington, S. (2012) Does personality matter for small business success? *South African Journal of Economic and Management Sciences* 15(4), 382–401.

Feldman, S. and Walsh, R. (1995) Feminist knowledge claims. Local knowledge and gender division of agricultural labor. *Rural Sociology* 60, 23–43.

Fernet, C., Torres, O., Austin, S. and St-Pierre, J. (2016) The psychological costs of owning and managing a SME: linking job stressors, occupational loneliness, and burnout. *Burnout Research* 3, 45–53.

Furnham, A. and Crump, J. (2015) The Myers–Briggs Type Indicator (MBTI) and promotion at work. *Psychology* 6, 1510–1515. DOI: 10.4236/psych.2015.612147.

Gardner, H. (1993) *Multiple Intelligence: The Theory and Practice*. Basic Books, New York.

Garforth, C. and Rehman, T. (2005) Review of Literature on Measuring Farmers' Values, Goals and Objectives. Project Report No. 2 in the 'Research to understand and model the behaviour and motivations of farmers responding to policy changes (England)'. School of Agriculture, Policy and Development, The University of Reading, Reading, UK.

Gasson, R. (1973) Goals and values of farmers. *Journal of Agricultural Economics* 24(3), 521–542.

Greig, B., Nuthall, P. and Old, K. (2017) *Results from a 2015 Survey of NZ farm Managers/Owners Covering Debt and Related Issues Designed to Explore the Impact of Debt*. Faculty of Agribusiness and Commerce Working Paper Series No. 18. Lincoln University, Christchurch, New Zealand. Available at: https://hdl.handle.net/10182/8339 (accessed 29 August 2024).

Greig, B., Nuthall, P. and Old, K. (2019) Farmers' characteristics' and the propensity to reduce net debt: the case for New Zealand (NZ) primary producers. *Agricultural Finance Review* 79(5), 614–632. DOI: 10.1108/AFR-02-2019-0020.

Greig, B., Nuthall, P. and Old, K. (2020) Exploring farmers' human characteristics as drivers of their anxiety. *Journal of Agromedicine* 25(1), 135–146. DOI: 10.1080/1059924X.2019.1656692.

Hamzah, M.I. and Othman, A.K. (2023) How do locus of control influence business and personal success? The mediating effects of entrepreneurial competency. *Frontiers in Psychology* 13, 958911. DOI: 10.3389/fpsyg.2022.958911.

Hanson, D.G., Parsons, R.L., Chess, W.J. and Balliet, K.L. (2002) Farm production analysis training for small farmers. *Journal of Extension* 40(4), 11.

Henry, M., Erwee, R. and Kong, E. (2013) Family business succession – trust and gender issues in family and non-family succession. In: Proceedings of the 13th Annual Conference of the European Academy of Management, Istanbul, Turkey, 26–29 June 2013. Available at: https://research.usq.edu.au/download/158aa5196e29f050026fa751bd7994e02967673d182cb0edd532b8e0f3cc209e/334629/Henry_Erwee_Kong_EURAM2013_PV.pdf (29 August 2024).

Higgs, M. (2001) Is there a relationship between the Myers–Briggs type indicator and emotional intelligence? *Journal of Managerial Psychology* 16(7), 509–533.

Hite, L.M. and McDonald, K.S. (1995) Gender issues in management development: Implications and research agenda. *Journal of Management Development* 14(4), 5–15.

Hobbs, D.J., Beal, G.M. and Bohlen, J.M. (1964) The Relation of Farm Operator Values and Attitudes to Their Economic Performance. Rural Sociology Report No. 33. Department of Economics and Sociology, Iowa State University, Ames, Iowa.

Hogarth, R. (2001) *Educating Intuition*. University of Chicago Press, Chicago, Illinois.

Hogarth, R. (2010) Intuition: a challenge for psychological research on decision making. *Psychological Inquiry* 21, 338–353.

Howard, W.H., Brinkman, G.L. and Lambert, R. (1997) Thinking styles and financial characteristics of selected Canadian farm managers. *Canadian Journal of Agricultural Economics* 45, 39–49.

Hughes, D.J., Kratsiotis, I.K., Niven, K. and Holman, D. (2020) Personality traits and emotion regulation: A targeted review and recommendations. *Emotion* 20(1), 63–67. DOI: 10.1037/emo0000644.

Izadi, Z., Ziyadin, S., Palazzo, M. and Sidhu, M. (2020) The evaluation of the impact of innovation management capability to organisational performance. *Qualitative Market Research* 23(4), 697–723.

Jackson-Smith, D., Trechter, D. and Splett, N. (2004) The contribution to financial management training and knowledge in dairy farm financial performance. *Review of Agricultural Economics* 26(1), 132–147.

John, O.P. and Gross, J.J. (2004) Health and unhealthy emotions regulation: personality processes, individual differences, and life span development. *Personality* 72(6), 1301–1333.

Johnsen, G.J. and McMahon, R.G.P. (2005) Owner–manager gender, financial performance and business growth amongst SMEs from Australia's Business Longitudinal Survey. *International Small Business Journal* 23(2), 115–142. DOI: 10.1177/0266242605050509.

Johnson, G.L., Halter, A.N., Jensen, H.R. and Thomas, D.W. (eds) (1961) *A Study of Managerial Processes of Midwestern Farmers*. Iowa State University Press, Ames, Iowa.

Jose, H.D. and Crumly, J.A. (1993) Psychological type of farm/ranch operators: Relationship to financial measures. *Review of Agricultural Economics* 15, 121–132.

Judge, T.A. and Ilies, R. (2002) Relationship of personality to performance motivation: A meta-analytic review. *Journal of Applied Psychology* 87(4), 797–807.

Judge, T.A., Simon, L.S., Hurst, C. and Kelley, K. (2014) What I experienced yesterday is who I am today: relationship of work motivations and behaviors to within-individual variation in the five-factor model of personality. *Journal of Applied Psychology* 99(2), 199–221. DOI: 10.1037/a0034485.

Jurik, N.C., Křížková, A., Pospíšilová, M. and Cavender, G. (2019) Blending, credit, context: Doing business, family and gender in Czech and US copreneurships. *International Small Business Journal* 37(4), 317–342. DOI: 10.1177/0266242618825260.

Kaine, G., Sandall, J. and Bewsell, D. (2004) Personality and strategy in agriculture. In: *Proceedings of the 20th AIAEE Annual Conference,* Dublin, Ireland. Association for International Agricultural and Extension Education, pp. 790–801.

Kanfer, R. and Ackerman, P. (2000) Individual differences in work motivation: Further explorations of a trait framework. *Applied Psychology* 49(3), 470–482.

Kayes, D.C. (2002) Experiential learning and its critics. Preserving the role of experience in management learning and education. *Academy of Management Learning and Education* 1(2), 137–149.

Keirsey, D. (1998) *Please Understand Me II: Temperament, Character, Intelligence*. Prometheus Book Co., Del Mar, California.

Kelly, E.L. and Conley, J.J. (1987) Personality and compatibility: A prospective analysis of marital stability and marital satisfaction. *Journal of Personality and Social Psychology* 52(1), 27–40.

Kelly, G.A. (1992 reprint) *The Psychology of Personal Constructs:* Volume One. Theory and Personality. Routledge, London.

Kilpatrick, S. (1997) Education and training: impacts on profitability in agriculture. *Australian and New Zealand Journal of Vocational Education Research* 5(2), 11–36.

Kilpatrick, S. (1999) Education and Training: Impacts on Farm Management Practice. Centre for Research and Learning in Regional Australia, University of Tasmania, Launceston, Australia.

Kim, J. and Zepeda, L. (2004) When the work is never done: Time allocation in US family farm households. *Feminist Economist* 10(1), 115–139.

Klein, G., Pliske, R., Crandall, B. and Woods, D. (2005) Problem detection. *Cognition, Technology and Work* 7, 14–28.

Kolb, A.Y. and Kolb, D.A. (2005) *Kolb Learning Style Inventory – Version 3.1. 2005 Technical Specifications*. HayGroup, Boston, Massachusetts.

Kolb, A.Y. and Kolb, D.A. (2013) *The Kolb Learning Style Inventory – Version 4.0. A Comprehensive Guide to the Theory, Psychometrics, Research on Validity and Educational Applications*. Experience Based Learning Systems, Inc., Kaunakakai, Hawaii.

Kolb, D.A., Rubin, I.M. and McIntyre, J.M. (1974) *Organizational Psychology: A Book of Readings*. Prentice-Hall, Englewood Cliffs, New Jersey.

Kolb, D.A., Rubin, I.M. and McIntyre, J.M. (1984) *Organizational Psychology: An Experiential Approach to Organizational Behavior*, 4th edn. Prentice-Hall, Englewood Cliffs, New Jersey.

Kwong, C., Jones-Evans, D. and Thompson, P. (2012) Differences in perceptions of access to finance between potential male and female entrepreneurs: Evidence from the UK. *International Journal of Entrepreneurial Behavior and Research* 18(1), 75–97.

Le Breton-Miller, I., Miller, D. and Steir, L. (2004) Toward an integrative model of effective FOB succession. *Entrepreneurship Theory and Practice* 28(4), 305–328.
Lerner, J., Li, Y., Valdesolo, P. and Kassam, K. (2015) Emotion and decision making. *Annual Review of Psychology* 66, 799–823.
Lobley, M., Baker, J. and Whitehead, I. (2012) Farm succession and retirement: Some international comparisons. *Journal of Agriculture, Food Systems and Community Development* 1, 49–64.
Matthews, G. and Deary, I.J. (1998) *Personality Traits*. Cambridge University Press, Cambridge.
McCall, M., Lombardo, M. and Morrison, A. (1988) *The Lessons of Experience: How Successful Executives Develop on the Job*. Lexington Press, Lexington, Massachusetts.
McCown, R., Carberry, P., Dalgliesh, N., Foale, M. and Hochman, Z. (2012) Farmers use intuition to reinvent analytic decision support for managing seasonal climatic variability. *Agricultural Systems* 106, 33–45.
McElwee, G. (2006) Farmers as entrepreneurs: developing competitive skills. *Journal of Developmental Entrepreneurship* 11, 187–206.
Menz, K.M. and Longworth, J.W. (1976) An integrated approach to farm management education. *American Journal of Agricultural Economics* 58(3), 551–556.
Mínguez-Vera, A. and Martin, A. (2011) Gender and management on Spanish SMEs: An empirical analysis. *The International Journal of Human Resource Management* 22(14), 2852–2873. DOI: 10.1080/09585192.2011.599948.
Minniti, M. (2009) Gender issues in entrepreneurship. *Foundations and Trends in Entrepreneurship* 5(7–8), 497–621. DOI: 10.1561/0300000021.
Montequin, V.R., Nieto, A.G., Ortega, F. and Villanueva, J. (2015) Managerial style profiles of successful project managers: a survey. *Procedia Computer Science* 64, 55–62.
Moon, S.J. (2021) Investigating beliefs, attitudes, and intentions regarding green restaurant patronage: an application of the extended theory of planned behavior with moderating effects of gender and age. *International Journal of Hospitality Management* 92, 102727. DOI: 10.1016/j.ijhm.2020.102727.
Muggen, G. (1969) Human factors in farm management: A review of the literature. *World Agricultural Economics and Rural Sociology Abstracts* 11, 1–11.
Myszkowski, N., Storme, M., Davila, A. and Lubart, T. (2015) Managerial creative problem solving and the big five personality traits: Distinguishing divergent and convergent abilities. *Journal of Management Development* 34(6), 674–684.
Nielsen, A.H. (1974) A management game for use in teaching farm management. *European Review of Agricultural Economics* 2(3), 293–306.
Niler, A.A., Asencio, R. and DeChurch, L.A. (2020) Solidarity in STEM: How gender composition affects women's experience in work teams. *Sex Roles* 82, 142–154. DOI: 10.1007/s11199-019-01046-8.
Nuthall, G. (2007) *The Hidden Lives of Learners*. NCER Press, Wellington.
Nuthall, G.A. and Alton-Lee, A. (1993) Predicting learning from student experience of teaching: A theory of student knowledge construction in classrooms. *American Educational Research Journal* 30(4), 799–840.
Nuthall, P. (2019) A review of the intuition literature relative to a recent quantitative study of the determinants of farmers' intuition. *International Journal of Agricultural Management* 8(1), 31–40. DOI: 10.5836/ijam/2019-08-31.
Nuthall, P. and Benbow, C. (1999) Computer System Uptake and Use on New Zealand Farms. Research Report 99/01. Farm and Horticultural Management Group, Lincoln University, Christchurch, New Zealand.
Nuthall, P.L. (2002) Managerial Competencies in Primary Production and The View of a Sample of New Zealand Farmers. FHMG Research Report 04/2002. Farm and Horticultural Management Group, Lincoln University, Christchurch, New Zealand.
Nuthall, P.L. (2006) Determining the important management skill competencies: The case of family farm business in New Zealand. *Agricultural Systems* 88, 429–450.
Nuthall, P.L. (2009) Modelling the origins of managerial ability in agricultural production. *Australian Journal of Agricultural and Resource Economics* 53, 413–436.
Nuthall, P.L. (2010) *Farm Business Management: The Core Skills*. CAB International, Wallingford, UK.
Nuthall, P.L. (2012) The intuitive world of farmers – the case of grazing management systems and experts. *Agricultural Systems* 107, 65–73.
Nuthall, P.L. (2013) *Factors Associated with the Development of Effective Experience in Excellent Managerial Ability*. Faculty of Commerce Research Report No. 1. Lincoln University, Christchurch, New Zealand. Available at: https://hdl.handle.net/10182/5526 (accessed 29 August 2024).

Nuthall, P.L. (2016) *The Intuitive Farmer: Inspiring Management Success*. 5M Publishing, Sheffield, UK.
Nuthall, P.L. (2017) *Farm Business Management: The Fundamentals of Good Practice*. CAB International, Wallingford, UK.
Nuthall, P.L. (2022) Assessing the core variables of business managers' intuitive decision ability: A review and analysis. *Behavioral Sciences* 12, 409. DOI: 10.3390/bs12110409.
Nuthall, P.L. and Old, K.M. (2014) Report on a Succession and Governance Survey of a Random Stratified Sample of New Zealand Farms. Research Report 331. Agribusiness and Economics Research Unit, Lincoln University, Christchurch, New Zealand. Available at: https://hdl.handle.net/10182/6261 (accessed 29 August 2004).
Nuthall, P.L. and Old, K.M. (2016) Farm owners' reluctance to embrace family succession and the implications for extension: The case of family farms in New Zealand. *The Journal of Agricultural Education and Extension* 23, 39–60.
Nuthall, P.L. and Old, K.M. (2017) Will future land based food and fibre production be in family or corporate hands? An analysis of farm land ownership and governance considering farmer characteristics as choice drivers. The New Zealand case. *Land Use Policy* 63, 98–110.
Nuthall, P.L. and Old, K.M. (2018) Intuition, the farmers' primary decision process. A review and analysis. *Journal of Rural Studies* 58, 28–38.
Ohlmer, B. (2001) Analytic and intuitive decision making – Swedish farmers' behaviour in strategic problem solving. In: *Proceedings of the 3rd EFITA Conference,* Montepellier. European Federation for Information Technology in Agriculture, Food and the Environment, Paris, pp. 609–614.
Ohlmer, B. (2007) *The Need and Design of Computerized Farm Management Tools. Lessons Learned from a Swedish Case.* Working Paper Series 2007:5. Department of Economics, Swedish University of Agricultural Sciences, Uppsala, Sweden.
Ohlmer, B., Olson, K. and Brehmer, B. (1998) Understanding farmers' decision making processes and improving managerial assistance. *Agricultural Economics* 18, 273–290.
Pfister, H. and Bohm, G. (2008) The multiplicity of emotions: A framework of emotional functions in decision making. *Judgement and Decision Making* 3(1), 5–17.
Piva, E. and Rovelli, P. (2023) Applicant attraction to new ventures: The combined effects of entrepreneurs' physical attractiveness and gender. *Journal of Small Business Management* 62(5), 2523–2562. DOI: 10.1080/00472778.2023.2246044.
Polanyi, M. (1966) The logic of tacit inference. *Philosophy* 41(155), 1–18. DOI: 10.1017/S0031819100066110.
Proudfoot, J.G., Corr, P.J., Guest, D.E. and Dunn, G. (2009) Cognitive-behavioural training to change attributional style improves employee well-being, job satisfaction, productivity, and turnover. *Personality and Individual Differences* 46, 147–153.
Psacharopoulos, G. and Patrinos, H.A. (2004) Returns to investment in education: A further update. *Education Economics* 12(2), 111–134.
Puntaier, E., Zhu, T. and Hughes, P. (2022) Board diversity as strategic choice and why it should matter to SMEs. *International Journal of Entrepreneurial Behavior & Research*. Online version ahead of print. DOI: 10.1108/IJEBR-03-2022-0225.
Rah, J.H., Hasler, C., Painter, J. and Chapman-Novakofski, K. (2004) Applying the theory of planned behavior to women's behavioral attitudes on and consumption of soy products. *Journal of Nutrition Education and Behavior* 36, 238–244.
Rickson, S.T. and Daniels, P.L. (1999) Rural women and decision making: Women's role in resource management during rural restructuring. *Rural Sociology* 64(2), 234–250.
Robie, C., Brown, D. and Bly, P.R. (2008) Relationship between major personality traits and managerial performance: Moderating effects of derailing traits. *International Journal of Management* 25(1), 131–139.
Rosa, P., Hamilton, D., Carter, S. and Burns, H. (1992) The impact of gender on small business management: preliminary findings of a British study. *International Small Business Journal* 12(3), 25–32.
Rougour, C.W., Tripp, G., Huirne, R.B.M. and Renkema, J.A. (1998) How to define and study farmers' management capacity: Theory and use in agricultural economics. *Agricultural Economics* 18, 261–272.
Sambodo, L.A.A.T. and Nuthall, P.L. (2010) A behavioural approach to understanding semi-subsistence farmers' technology adoption decisions: The case of improved paddy–prawn system in Indonesia. *The Journal of Agricultural Education and Extension* 16, 111–129.
Samobodo, L.A.A.T. and Nuthall, P.L. (2009) *The Decision Making Processes of Semi-Commercial Farmers: A Case Study of Technology Adoption in Indonesia*. VDM, Saarbrucken, Germany.

Santos, F.J., Roomi, M.A. and Liñán, F. (2016) About gender differences and the social environment in the development of entrepreneurial intentions. *Journal of Small Business Management* 54(1), 49–66. DOI: 10.1111/jsbm.12129.

Sawer, B. (1974) *The Role of the Wife in Farm Decisions*. Rural Sociology Monograph No. 5. Adult Education Research Centre, University of British Columbia, Vancouver, Canada.

Schreibauer, E.C., Hippler, M., Burgess, S., Rieger, M.A. and Rind, E. (2020) Work-related psychosocial stress in small and medium-sized enterprises: An integrative review. *International Journal of Environmental Research and Public Health* 17(20), 7446. DOI: 10.3390/ijerph17207446.

Shoubaki, A.E., Laguir, I. and den Besten, M. (2020) Human capital and SME growth: The mediating role of reasons to start a business. *Small Business Economics* 54, 1107–1121.

Simo Kengne, B. (2016) Mixed-gender ownership and financial performance of SMEs in South Africa: A multidisciplinary analysis. *International Journal of Gender and Entrepreneurship* 8(2), 117–136. DOI: 10.1108/IJGE-10-2014-0040.

Sinclair, M. (ed.) (2014) *Handbook of Research Methods on Intuition*. Edward Elgar Publishing, Cheltenham, UK.

Singh, G. and Belwal, R. (2008) Entrepreneurship and SMEs in Ethiopia: evaluating the role, prospects and problems faced by women in this emergent sector. *Gender in Management* 23(2), 120–136. DOI: 10.1108/17542410810858321.

Sok, J., Hogeveen, H., Elbers, A.R.W. and Oude Lansink, A.G.J.M. (2016) Perceived risk and personality traits explaining heterogeneity in Dutch dairy farmers' beliefs about vaccinating against bluetongue. *Journal of Risk Research* 21, 562–578.

Solano, C., Leon, H., Perez, E., Tole, L., Fawcett, R.H. *et al.* (2006) Using farmer decision making profiles and managerial capacity as predictors of farm management performance in Costa Rican dairy farmers. *Agricultural Systems* 88, 395–428.

Sternberg, R.J. (1995) *In Search of the Human Mind*. Harcourt Brace, Fort Worth, Texas.

Sternberg, R.J., Wagner, R.K., Williams, W.M. and Horvath, J.A. (1995) Testing common sense. *The American Psychologist* 50(11), 912–927.

Stewart, W.H., Watson, W.E., Carland, J.C. and Carland, J.W. (1998) A proclivity for entrepreneurship: A comparison of entrepreneurs, small business owners, and corporate managers. *Journal of Business Venturing* 14(2), 189–214.

Strawser, J.A., Hechavarría, D.M. and Passerini, K. (2021) Gender and entrepreneurship: research frameworks, barriers and opportunities for women entrepreneurship worldwide. *Journal of Small Business Management* 59(Suppl. 1), S1–S15. DOI: 10.1080/00472778.2021.1965615.

Summer, D.A. and Lieby, J.D. (1987) An econometric analysis of the effect of human capital on size and growth among dairy farms. *American Journal of Agricultural Economics* 69, 465–470.

Szcześniak, M., Rodzeń, W., Malinowska, A. and Kroplewski, Z. (2020) Big five personality traits and gratitude: The role of emotional intelligence. *Psychology Research and Behavior Management* 13, 977–988. DOI: 10.2147/PRBM.S268643.

Taylor, S.E. and Armor, D.A. (1996) Positive illusions and coping with adversity. *Journal of Personality* 64(4), 873–898.

Taylor, S.E., Pham, L.B., Rivkin, I.D. and Armor, D.A. (1998) Harnessing the imagination. Mental stimulation, self-regulation, and coping. *The American Psychologist* 53(4), 429–439.

ten Raa, T. (2022) Activity analysis in production economics. In: Ray, S.C., Chambers, R.G. and Kumbhakar, S.C. (eds) *Handbook of Production Economics*. Springer, Singapore, pp. 471–482. DOI: 10.1007/978-981-10-3455-8_25.

Thomas, H.V., Lewis, G., Thomas, D.Rh., Salmon, R.L., Chalmers, R.M. *et al.* (2003) Mental health of British farmers. *Occupational and Environmental Medicine* 60, 181–186.

Tripp, G., Thijssen, G.J., Renkema, J.A. and Huirne, R.B.M. (2000) Measuring managerial efficiency: The case of commercial greenhouse growers. *Agricultural Economics* 27, 172–181.

Verbano, C. and Venturini, K. (2013) Managing risks in SMEs: A literature review and research agenda. *Journal of Technology Management & Innovation* 8(3), 186–197. DOI: 10.4067/S0718-27242013000400017.

Wagner, R. and Sternberg, R. (1987) Tacit knowledge in managerial success. *Journal of Business and Psychology* 1, 301–312.

Warren, R.D., White, J.K. and Fuller, W.A. (1974) An errors-in-variables analysis of managerial role performance. *Journal of the American Statistical Association* 69, 886–893.

Westbrooke, V. and Nuthall, P. (2017) Why small farms persist? The influence of farmer's characteristics on farm growth and development. The case of smaller dairy farmers in NZ. *Australian Journal of Agricultural and Resource Economics* 61, 663–684.
Willock, J., Deary, I.J., McGregor, M.M., Sutherland, A., Edwards-Jones, G. *et al.* (1999) Farmers' attitudes, objectives, behaviours, and personality traits: The Edinburgh Study of Decision Making on Farms. *Journal of Vocational Behavior* 54, 5–36.
Wilson, P., Hadley, D., Ramsden, S. and Kaltsas, I. (1998) Measuring and explaining technical efficiency in UK potato production. *Journal of Agricultural Economics* 49, 294–305.
Wolf de, P.L., Schoorlemmer, H.B., Smit, A.B. and de Lauwere, C.C. (2004) Analysis and development of entrepreneurship in agriculture. In: Bokeman, W. (ed.) *XV International Symposium on Horticultural Economics and Management. Acta Horticulturae* 655, 199–208.
Xayavong, V., Kingwell, R. and Islam, N. (2015) How training and innovation link to farm performance: A structural equation analysis. *Australian Journal of Agricultural and Resource Economics* 59, 1–16.
Xin Ding, D. (2014) The effect of experience, ownership and focus on productive efficiency: A longitudinal study of US hospitals. *Journal of Operations Management* 32, 1–14.
Yang, T., Hughes, K.D. and Zhao, W. (2021) Resource combination activities and new venture growth: exploring the role of effectuation, causation, and entrepreneurs' gender. *Journal of Small Business Management* 59(Suppl. 1), S73–S101. DOI: 10.1080/00472778.2020.1790292.
Young, A.J. and Walters, J.L. (2002) Relationship between DHI production values and Myers–Briggs type indicator as a measure of management ability. *Journal of Dairy Science* 85, 2046–2052.
Zeidner, M., Matthews, G. and Roberts, R.D. (2004) Emotional intelligence in the workplace. *Applied Psychology* 53(3), 371–390.
Zuckerman, M., Joireman, J., Kraft, M. and Kuhlman, D.M. (1999) Where do motivational and emotional traits fit within the three factor models of personality? *Personality and Individual Differences* 26(3), 487–504.

Index

Note:- page numbers in **bold** type refer to figures; page numbers in *italic* type refer to tables.